The Politics of Climate Change under President Obama

The last two decades have witnessed an ever-growing partisan divide in US politics over climate change and global warming. Significant elements in the Republican Party became openly hostile to the scientific evidence and, following the election of George W. Bush, legislative action at the federal level effectively ground to a halt. This opened up space at the state and local level to develop climate change policies, with cities such as Chicago, San Francisco and New York implementing a number of initiatives that brought real and substantive developments. The election of Barack Obama in 2008 seemed to open new possibilities for federal and global leadership once more and whilst the Obama administration has been criticised for a somewhat contradictory approach to the issue at times, there were nonetheless a number of significant policy developments.

Through a substantive and detailed analysis of the politics of climate change, this book places the evolution of US climate policy within broader debates on the nature of politics in the US and argues that there exists a latent potential, often obscured by the complexities of its political system, for America to act as a world leader on the issue.

This work will appeal particularly to students and scholars in American Politics, but will also prove useful to those in the fields of general politics, climate change, sustainability and environmental studies.

Hugh Atkinson is a tutor in politics and sustainability at London South Bank University. He is a distinguished research fellow at the Schumacher Institute for Sustainable Solutions, Bristol. His main research specialism is in the politics of climate change.

The Politics of Climate Change under President Obama

Hugh Atkinson

Routledge
Taylor & Francis Group

LONDON AND NEW YORK

First published 2018
by Routledge
2 Park Square, Milton Park, Abingdon, Oxon OX14 4RN

and by Routledge
605 Third Avenue, New York, NY 10017

First issued in paperback 2021

Routledge is an imprint of the Taylor & Francis Group, an informa business

Publisher's Note
The publisher has gone to great lengths to ensure the quality of this reprint
but points out that some imperfections in the original copies may be
apparent.

British Library Cataloguing in Publication Data
A catalogue record for this book is available from the British Library

Library of Congress Cataloging in Publication Data
A catalog record for this book has been requested

Typeset in Times New Roman
by Wearset Ltd, Boldon, Tyne and Wear

ISBN 13: 978-1-03-224206-4 (pbk)
ISBN 13: 978-1-4724-4662-6 (hbk)

DOI: 10.4324/9781315598642

Contents

Illustrations

Figures

Tables

Preface

Like many others of my generation, in my childhood I was a long distance observer of America. My first impressions were forged in the late 1950s and early 1960s with the likes of US TV shows such as Mr Ed, I Love Lucy, the Lone Ranger, and the irrepressible Beverley Hillbillies.

On a much more sombre note, I still recall with vivid clarity that dark November day in 1961 when as an eight year old I heard the terrible news of the assassination of JFK. I have remained a keen observer of American politics ever since.

My special thanks to Rob Sorsby, commissioning editor at Taylor and Francis. His support and encouragement have been central in getting this book written. I would also like to thank my wife and academic colleague, Professor Ros Wade for her undying support in the writing of this book. She not only acted as an unpaid consultant and researcher but put up with my moods and complaints. Her support and love was pivotal to completing this book. A special mention too for Tim Berners-Lee, the inventor of the world wide web which has done so much to open up academic research and debate.

I would also like to thank the gorgeous village of Wincle in East Cheshire. I spent two weeks there in a lovely cottage. My time there gave me inspiration in the final stages of writing this book. The peace and quiet, the beautiful walks by the River Dane, and a few pints of ale at the local village pub proved the perfect combination. I strongly recommend it to anyone writing a book.

Abbreviations and acronyms

ACCCE	American Coalition for Clean Coal Electricity
ACEEE	American Council for an Energy Efficient Economy
ANWR	Arctic National Wildlife Refuge
ARRA	American Recovery and Investment Act
C2ES	Centre for Climate and Energy Solutions
CAP	climate action plan
CCP	comprehensive conservation plan
CCPI	Climate Change Performance Index
CCS	carbon capture and storage
CCTI	Climate Change Technology Initiative
CLC	Climate Leadership Council
COP21	UN Conference of the Parties on Climate Change
CPA	Climate Protection Agreement
CCPL	community climate plan
CPP	Clean Power Plan
CSD	Council on Sustainable Development
DEP	Department of Environmental Protection
DOE	US Department of Energy
DOI	US Department of Industry
EERS	energy efficiency resource standard
EIA	US Energy Information Administration
EJF	Environmental Justice Foundation
EPA	Environmental Protection Agency
GCF	Green Climate Fund
GE	General Electric
HFCs	hydrofluorocarbons
ICLEI	International Council for Local Environmental Initiatives
ICS	investment court system
IEA	International Energy Agency
IIED	International Institute for Environment and Development
INDCs	intended nationally determined contributions
IPCC	United Nations Intergovernmental Panel on Climate Change
ISDS	investor state dispute settlement

MMS	Minerals Management Service
MW	megawatts
NADA	National Automobile Dealers Association
NAFTA	North America Free Trade Area
NCADAC	National Climate Assessment and Development Advisory Committee
NCAR	National Centre for Atmospheric Research
NIC	National Intelligence Council
NOAA	National Oceanic and Atmospheric Administration
NRDC	National Resources Defence Council
NRECA	National Rural Electric Co-operative Association
NSIDC	National Snow and Data Ice Centre
OCS	Outer Continental Shelf
OPEC	oil producing exporting countries
PEJ	Project for Excellence in Journalism
PlaNYC	New York City Plan
PPC	Program for Public Consultation
PV	solar photovoltaic power
RGGI	Regional Greenhouse Gas Initiative
RPS	renewable portfolio standard
SCI	Sustainable Cities Index
SDGs	sustainable development goals
SIPRI	Stockholm International Peace Research Institute
TPP	Trans-Pacific Partnership
TRI	Toxic Release Inventory Programme
TTIP	Transatlantic Trade and Investment Partnership
UCS	Union of Concerned Scientists
UNCED	1992 UN Conference on Environment and Development
UNFCCC	United Nations Framework Convention on Climate Change
USGS	United States Geological Survey
WCI	Western climate initiative
WEO	World Energy Outlook
WMO	World Meteorological Organisation

Introduction

The key focus of this book is the eight years of the Obama Presidency, set within the context of a political culture that has developed over the last two decades; a political culture that has seen a significant increase in political polarisation. It addresses the issue of climate change and the policy response of the US towards this ever-growing policy challenge.

We are now living in what has been described as the 'anthropocene' era. The term was originally coined by the ecologist Eugene F. Stormer, but popularised by the atmospheric chemist and Nobel laureate, Paul Crutzen. The argument for a new era is straightforward: the impact of human behaviour over a consolidated period of time has been so significant as to constitute a distinct geological epoch. This is especially so with the impact of climate change on planet earth. The impact of climate change across the globe is there for all to see. The ice in the Arctic is receding at an alarming rate, sea levels are rising, and the world is experiencing a series of extreme weather events.

The 2006 Stern Report (Stern 2006) spelled out in clear terms the environmental, social and economic consequences if politicians, decision makers and the public at large do not make the right choices now and in the years ahead. Stern warned that delicate eco systems were under threat. He also noted rising sea levels were significantly increasing the risk of flooding, putting at risk water supplies to one sixth of the population of the planet. In the US, more and more evidence points to the impact of increasing storm surges, floods and droughts on the lives of millions of Americans. California's Environmental Protection Agency (EPA) reports that lakes are warming, sea levels are rising, and wild fires are on the increase in the state due to the impact of climate change. Recent data published by the United Nations (UN) intergovernmental panel on climate change (IPCC) shows that greenhouse gases continue to rise, with the likelihood of global temperature increases of 2°C (3.6°F) from pre industrial times.

As far back as 1988, the US climate scientist James Hansen presented evidence to the US Congress on the link between climate change and the increase in greenhouse gases in the atmosphere. Yet, there is a widespread pessimistic view about the ability and commitment of the US to tackle climate change. Sussman and Daynes, for example, talk of the 'concern that there is a general lack of interest within the United States regarding the global consensus on climate change' (Sussman and Daynes 2013: 1). In a similar vein, Rabe notes that 'The US has been widely condemned for its repeated failure to forge bold national strategies to address climate change' (Rabe 2015: 55). Part of the explanation for this lies in the fact that climate change is a complex and wicked problem that presents fundamental challenges for both politicians and voters about the way we live our lives now and in the future. But these are challenges that need to be met if we are to safeguard the planet for current and future generations.

In the context of the US, this book will argue that the record of the eight years of the Obama Presidency with regard to climate policy marked a step change from both the tone and the policy actions of the previous administration of the Republican George W. Bush. Within three months of taking office, the Bush administration announced that it would not be implementing the 1997 Kyoto Protocol. The Kyoto Protocol was an international agreement on climate change which set targets for the reduction of greenhouse gases. This was just one of a number of policies which sought to row back action on climate change and environmental protection, to the anger of the environmental lobby.

The Democrat, Barack Obama, took up the Presidential reigns of office in January 2009. In contrast to Bush, his election was welcomed by environmental groups. Addressing the UN General Assembly in September 2009, Obama spoke of the responsibility of the richer nations, including the US, to take the lead in the fight on climate change. There followed a number of important climate change policy initiatives, but faced with an increasingly hostile Republican-dominated Congress, there was limited legislative action. Instead, Obama had to pursue his policy objectives through a series of executive actions and orders.

The book is divided into six chapters. Chapter 1 is divided into three sections. First the policy challenge of climate change is set in its broader context, with a focus on the nature of climate change, its causes and its environmental impact. Second there will be an analysis of the so-called 'golden age' of US environmental policy. Third, there is a broad survey of the politics of climate change in the US over the last two decades. The chapter sets the context for a more substantive and detailed analysis of the politics of climate change in the chapters that follow. There is an analysis of the environmental and climate change record of Presidents Bill Clinton and George W. Bush.

Chapter 2 focuses on various aspects of President Obama's climate change agenda. The chapter will analyse a number of key policy initiatives. These include the Clean Power Plan (CPP) and the contribution of the Obama administration to the 2015 UN Paris COP21 climate change conference. An important focus of the chapter will be the increasing political polarisation around the issue of climate change, with Congressional Republicans opposing the Presidential agenda at every turn. As a consequence, Obama had to increasingly rely on executive actions to develop policy on climate change.

Chapter 3 will argue that the climate change policy process in the US is a very crowded arena with vested interests on all sides seeking to push their particular agendas. There will be an analysis of the power and influence of pressure groups, with a focus on theoretical models which seek to explain the locus and impact of power. The role of the courts in the policy process will be examined. There will also be an analysis of the role of the media and public opinion in framing the debate around climate.

Chapter 4 will focus on energy policy and the energy mix in the US and the relationship to action on climate change. Energy policy is central to any effective strategy to tackle climate change; in particular moving away from a reliance on fossil fuels such as coal, gas and oil. There will be an analysis of the policy approach to renewable energy sources such as solar power. The chapter will also note some of the apparent conflicts and contradictions within US energy policy. There will also be a focus on the current boom in shale gas in the US.

Chapter 5 will argue that the relative lack of action on climate change at the federal level of government in the US has opened up a policy space which a number of states and cities have sought to fill. While the administration of George W. Bush rejected the 1997 Kyoto Protocol, some states and cities adopted a number of the Kyoto principles in their climate action plans. However, there is much variability in the action at the state and city level, with areas of Democratic support much more likely to take action on climate change than their Republican counterparts.

Finally, Chapter 6 will look at the role of the US as world leader in tackling climate change and building a more sustainable world. It will argue that, although the US has historically contributed significantly to the problems that people and planet face with regard to climate change, it also has both the potential, and indeed the obligation, to be at the forefront in developing strategies to tackle climate change and build a more sustainable world. Indeed without effective leadership and participation by the US, the world is most unlikely to deal effectively with the challenges that climate change presents.

References

Rabe, B. (2015) 'A new era in states' climate policies' in Wolinssky-Nahmias, Y. (ed.), *Climate Change Politics: US Policies and Civic Action*, Los Angeles: Sage.

Stern, N. (2006) *Stern Review on Economics of Climate Change*, London: HMSO.

Sussman, G. and Daynes, B. (2013) *US Politics and Climate Change: Science Confronts Policy*, Boulder, CO: Lynne Rienner.

1 Environmental policy and climate change

Two decades of change, consensus and conflict

Over recent times there has developed an understandable view of the United States of America as a country with only a limited commitment to the fight against climate change and with limited engagement with the broader sustainability agenda. Indeed as Reich argues, 'As a nation the United States seems incapable of doing what is required to reduce climate change' (Reich 2008: 5). The failure of the US Congress to ratify the 1997 Kyoto Protocol on the reduction of greenhouse gases and the record of the George W. Bush Presidency certainly seems to back up such a perception. However, this chapter will argue that the actual picture is far more nuanced and complex.

The last two decades have seen an ever-growing partisan divide with the US political system. The debate between the two main political parties, the Democrats and Republicans, has become ever more rancorous and toxic. This intensified with the election of Barack Obama back in 2008. All too often policy debates about the environment and tackling climate change in the US have been drowned out by the white noise of an increasingly partisan and hysterical political culture. Add in the constitutional exigency of the 'separation of powers' and the result has been policy gridlock across a range of issues. Action on tackling climate change has not been immune from this. Leading politicians, specifically conservative Republicans and Tea Party supporters, 'pour scorn on the very existence of a climate change problem, ignore or subvert the science, and argue that the whole thing is a put up job designed to increase big government and undermine American values' (Atkinson 2015: 89).

Yet it was not always like this. In the period from 1964 to 1980, during what has been described as the 'golden age' of US environmental policy, a wide range of legislation was passed and policies enacted. These included support for the eco system, the protection of wildlife, and measures to protect water and air quality. There was a broad consensus across the political spectrum on the desirability of such policies. Nothing better illustrates this consensus than the creation of the US EPA. It was supported by a Democratic dominated Congress and signed into law by Republican President Richard Nixon.

The chapter is divided into three sections. First, the policy challenge of climate change is set in its broader context. Second, there will be an analysis of the 'golden age' of environmental policy. Third, there is a broad survey of the politics of climate change in the US over the last two decades. This chapter will set the scene and provide the context for a more detailed analysis of the politics of climate change in the US in subsequent chapters.

The challenge of climate change: evidence and debate

We are now living in what has been described as the anthropocene era. This is an argument that the impact of human behaviour on planet earth over a sustained period of time has been so significant as to constitute a new geological epoch. This is no more evident than in the challenge of climate change. As Bierman has cogently argued, 'Humans now influence all biological and physical systems of the planet. Almost no species, land area, or part of the oceans has remained unaffected by the expansion of the human species' (Bierman 2014: viii).

There is now an overwhelming consensus in the global scientific community that climate change is real and that it is the result of human action in the shape of the extensive use of fossils fuels such as oil, coal and gas since the start of the Industrial Revolution. This has resulted in the release of large amounts of carbon dioxide gas into the earth's atmosphere. Carbon dioxide acts as a 'greenhouse' gas, trapping additional heat from the sun in the earth's atmosphere. This is referred to as the 'greenhouse effect'. As a consequence, the earth's temperature is rising steadily but inexorably with potentially grave and unpredictable consequences for both planet and people. Tackling climate change has taken on increased saliency over the last few years, but has been evident in policy circles for 30 years or more.

For example, on 23 June 1988 the American scientist James Hansen presented evidence to the US Senate. He told the assembled senators that he was 99 per cent certain that the record temperatures that year in the US were not the result of natural variations but were the consequence of growing concentrations of greenhouse gases in the earth's atmosphere. 'It is time to stop waffling so much,' argued Hansen, 'and say that the evidence is pretty strong that the greenhouse effect is here' (Worldwatch Institute 2009: 6). In the 28 years since Hansen gave his testimony that evidence has grown stronger.

The UN IPCC is made up of 200 of the world's leading climate scientists. Its stated position is unequivocal. Climate change and global warming are a clear and present threat and they are caused by the vast amounts of greenhouse gases (principally in the form of carbon dioxide) that we as human beings have been pumping into the world's atmosphere since the dawn of the Industrial Revolution by our continuing use of fossil fuels. As Grover and Peschek argue, 'The continuous burning of fossil fuels portends a global climate catastrophe' (Grover and Peschek 2014: 149). Figure 1.1 shows the amount of greenhouse gases being pumped into the atmosphere from 1900 to 2015 as measured in parts per million (ppm). It illustrates a clear and significant upward trend in such emissions.

According to data from the world meteorological organisation, in 2015 global greenhouse gas emissions averaged across the year reached the milestone of 400ppm for the first time. The US National Oceanic and Atmospheric Administration (NOAA) greenhouse gas monitoring station at Mauna Loa, Hawaii,

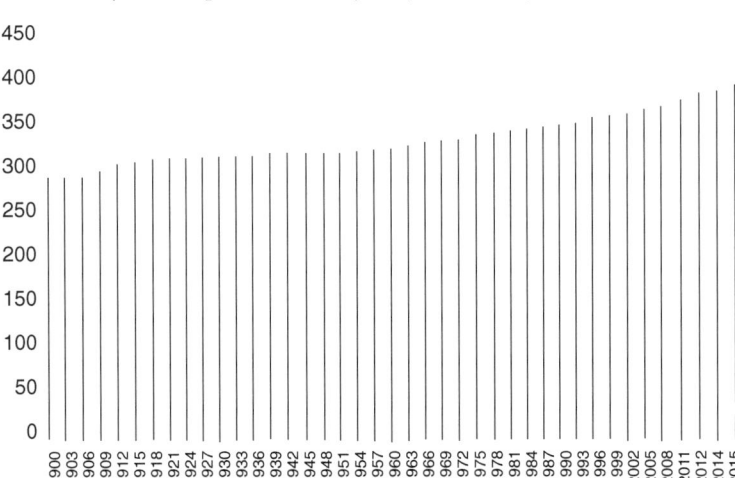

Atmospheric CO$_2$ concentration, parts per million, by volume, 1900–2015

Figure 1.1 Atmospheric greenhouse gases 1900–2015

Source: Base data compiled by Earth Policy Institute from Intergovernmental Panel on Climate Change (IPCC) and National Oceanic and Atmospheric Administration (NOAA), plus data from World Meteorological.

predicts that such concentrations are here to stay for the foreseeable future and will not drop below this level for many generations. The world has not seen such concentrations of greenhouse gases in the atmosphere for several million years since a time when the Arctic was ice-free, and sea levels were 40 metres higher than today.

The consequence of this continued rise in greenhouse gas emissions has been a steady but inexorable rise of global surface temperatures. Figure 1.2 illustrates the changes in the temperature of the earth's atmosphere in degrees Fahrenheit from 1900 to 2014. Of course there are variations from year to year, but there is a clear upward trend. It is important to note that 20 of the hottest 21 years since records began in 1860 have occurred in the last 25 years.

The latest evidence from the IPCC shows that global greenhouse gas emissions continue to rise, with the possibility of global temperature increases above 2°C within the next 20 to 30 years (IPCC 2013). The 2°C figure (equivalent to 3.6°F) is significant, with many scientists now of the view that any increase beyond this would take us into unchartered territory, with the world experiencing more and more severe and unusual weather events. In spite of the clear evidence of human induced climate change there is a significant climate change denial lobby, backed by big oil, which seeks to undermine the science at every turn. This lobby is particularly prevalent in the US with bodies such as the Ayn Rand Institute, the Cato Institute and the Heritage Foundation. I shall return to this theme in subsequent chapters.

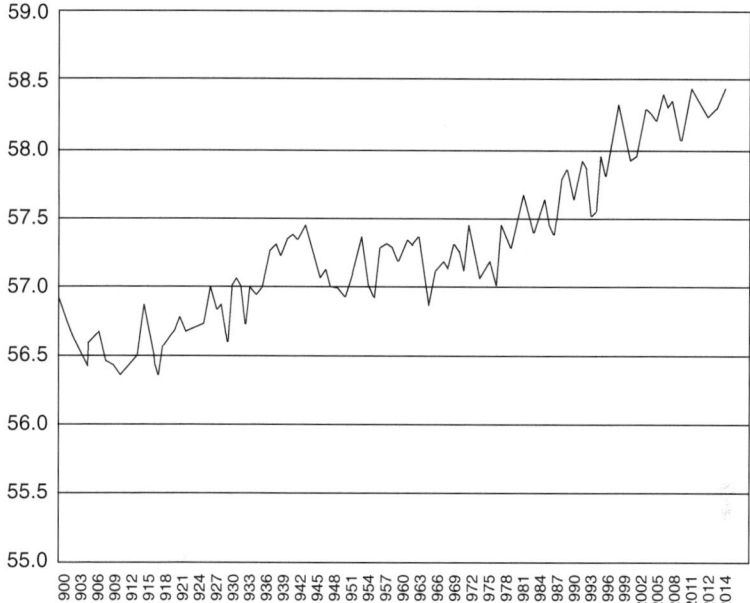

Figure 1.2 Average global surface temperatures 1900–2014

Source: Base data compiled by Earth Policy Institute (2014) from National Aeronautics and Space Administration (NASA) and Goddard Institute for Space Studies (GISS).

In its 2011 World Energy Outlook (WEO), the International Energy Authority (IEA), which has traditionally adopted a somewhat conservative approach to climate change, stated that the world was locking itself into an unsustainable energy future. The WEO has set out what it calls a '450 scenario' which traces an energy path consistent with meeting the globally agreed goal of limiting the temperature rise to 2°C (3.6°F). 80 per cent of energy related emissions are already locked in by existing infrastructure, such as power stations and factories. 'Without further action by 2017,' the IEA concludes, 'the energy related infrastructure then in place would generate all the CO_2 emissions allowed in (the) 450 scenario up to 2035' (IEA 2011).

Who is to blame?

So who are the biggest culprits when it comes to greenhouse gas emissions? Historically the US has been the largest contributor to greenhouse gas emissions (Friedrich and Damasson 2014). In 2014, China (with a population of 1.3 billion) was the largest emitter of greenhouse gas emissions, accounting for 30 per cent of total emissions. The US (with a population some 310 million) was second, accounting for some 15 per cent of emissions. The European Union accounted for 10 per cent of total emissions (PBL Netherlands 2015) This places a major responsibility on the US (alongside China and the European Union) to show political leadership in tackling climate both at home and on the global level. It is an argument I shall return throughout this book but in particular in Chapter 6.

The impact of climate change

It has been argued that

> Like a distant tsunami that is only a few metres high in the deep ocean but rises dramatically as it reaches shallow coastal waters, the great wave of climate change has snuck upon people – and is now beginning to bite.
>
> (Flavin and Engelman 2009: 6)

Across the globe evidence is mounting of the impact of climate change. As Bonds has argued, 'Global warming is transforming the Arctic' (Bonds 2016). Sea ice in the Arctic Sea has shrunk to the smallest level ever recorded based on data from the National Snow and Ice Data Center (NSIDC) in Colorado. Satellite images from 2012 show that a rapid melt has reduced the area of frozen sea to less than 3.3 million square kilometres, less than half that of 40 years ago. In the period from 2010 to 2012 the volume of Arctic ice fell by 14 per cent. However, it is only fair to point out that ice levels in the Arctic did stage a revival in 2013 with the volume of ice going up by 41 per cent. What is the explanation for this? As Carrington points out, this revival was the result of cooler temperatures in 2013 (Carrington 2017). So does this data undermine the case against climate change and strengthen the argument of the climate change denial lobby? Professor Rachel Tilling of University College London who led the Arctic study commented that while the upturn in Arctic ice volume was surprising, she rejected the idea of a wider recovery of the ice cap, arguing that climate change was still driving average temperatures up despite significant variations from year to year (Carrington 2017).

Indeed data published by the US National Snow and Ice Data Centre (NSIDC) in March shows that ice in the Arctic has fallen to a new winter time low; a consequence of unusually high temperatures in the polar regions. The ice cap grows during the winter months and usually reaches its maximum in early March. However, the maximum for March 2017 was 14.4 square kilometres lower than any year in the 38 years of satellite tracking. For Mark Serreze, the director of NSIDC, such developments are unprecedented. 'I have been looking at Arctic weather patterns for 35 years' said Serreze 'and we have never seen anything close to what we've experienced these past two winters' (Carrington 2017). 2017 is the third year in a row that the Arctic's winter ice has set a new low. Such major loss of ice poses significant threats to the planet with rising sea levels and more extremes of weather.

There are other pressures on the Arctic. It is estimated that the region holds up to 25 per cent of the world's remaining oil and gas reserves. The delicate eco system of the Arctic is coming under increasing threat as the melting ice eases the logistical difficulties of extracting oil. This presents commercial opportunities for oil corporations with further consequences for the Arctic and the planet generally as more and more greenhouse gases are pumped into the atmosphere. Late on in his Presidency, Barack Obama used his executive authority to introduce a five year moratorium on drilling in the US-controlled Arctic region. What the election of Donald Trump as President means for the region time alone will tell.

The 2006 Stern Report warned of the severe environmental, economic and social consequences of climate change with increasing sea levels increasing the risk of flooding and threatening water supplies to one sixth of the world's population. Delicate eco systems were coming under threat. Stern argued that urgent action was needed (Stern 2006: 56). Over a decade later the situation is even more grave and urgent. The physical impact of climate change is becoming more and more evident. Rising sea levels threaten the livelihoods of millions of people with all that means for poverty, social justice and political and social instability.

In the specific context of the US, evidence suggests that climate change will lead to rising sea levels which will threaten many coastal cities. Data also points to a loss of between 30 to 70 per cent of the snow pack in the Rocky Mountains; highly significant as it provides much of the water for the American west (Gertner 2007). It is forecast that rainfall in the west of the US will decrease by 20 per cent between 2040 and 2060 (Revkin 2008). Indeed there has been a five year drought in the state of California, broken only in early 2017 with storms and torrential downpours. In addition, climate change has reduced the Sierra snowpack, a vital stockpile of fresh water for California. These are major environmental challenges which need a political and policy response.

The US National Climate Assessment Report published in 2013 showed the impact of increasing storm surges, floods and droughts on the lives of Americans (NCADAC 2013). Research published by the California EPA shows that lakes are warming, sea levels are rising, and wildfires are spreading as the impact of climate takes hold throughout the state.

A recent report by the NOAA shows that 2012 was the warmest year on record for the contiguous US. The average temperature for the year was 55.3°F, a full degree warmer than the previous record year, 1998, and 3.3 degrees above the entire average of the twentieth century (NOAA 2013).

Furthermore, research from the NOAA shows that 2016 was the second warmest year on record for the contiguous US (NOAA 2017a). However, what is particularly striking about 2016 is how widespread the warming was. 'The breadth of the 2016 warmth is unparalleled in the nation's climate history,' explains the NOAA 'no other year had as many states (in the US) breaking or close to breaking their warmest annual average temperature' (Romm 2017). For the world's leading climatologist, Michael Mann, '2016 was an exclamation point, another record warm year in a record warm decade, filled with unprecedented, increasingly devastating extreme weather events'. He went on, 'It was mother nature's warning' to the climate change deniers in the Trump administration (Romm 2017).

Furthermore, data from the NOAA shows that the 2016 globally averaged surface temperature ended as the highest since records began in 1860. This marked three consecutive years of record warmth for the globe. These findings were backed up by separate research by NASA (NOAA 2017b). Furthermore, 14 out of the last 15 years have been the hottest on record.

The national intelligence council (NIC), a government advisory agency, looks at the strategic challenges facing the US and the world. In its 2016 report, it argues that 'climate change, environmental, and health issues will demand immediate attention'. It talks of 'a range of global hazards' which 'pose imminent and longer term threats that will require a collective response' (NIC 2017). These include more extreme weather, water and soil stress, food insecurity and increasing poverty. The report argues that rising sea levels, brought about by glacial melt such as the situation in the Arctic, will have a big impact on people and planet. The report warns that tensions over climate change will continue to grow.

As evidence of the challenge facing the world, leading climate scientists have warned that the Earth is coming very close to breaking through a 1.5°C (2.7°F) increase above pre-industrial levels. The decision to try to limit warming to these levels was one of the important outcomes of the United Nations Framework Convention on Climate Change (UNFCCC) agreed in Paris in December 2015. The agreement, more commonly known as COP21, is central to attempts to prevent some of the worst effects of climate change in the future such as the melting of the polar ice caps. However, Professor Chris Field, an academic at Stanford University and co chair of the IPCC working group on adaptation to climate change argues that 'the 1.5°C goal now looks impossible or at the very least, a very, very difficult task' (McKie 2016).

The golden age

As I have already noted, the current political debate in the US about climate change and the environment is highly conflicted, being marked by rancour and open hostility. Yet this has not always been the case. Though the issue of climate change itself was still to reach the policy agenda, on the broader issues of environmental policy there was a broad consensus across party lines during the period from 1964 to 1980 on the necessary legislative action. This period has often been referred to as the 'golden age' of environmental policy' (Bailey 2010: 168; Klyza and Sousa 2008: 1). The US Congress enacted 22 major laws in this period. One of the major achievements was the creation of the EPA, an organisation which today is at the receiving end of a major degree of opprobrium from large sections of the Republican Party, the Tea Party movement and big oil. The EPA was set up by Congress and the White House in response 'to the growing public demand for cleaner water, air and land' (Storey 2010: 342). Conceived in the context of increasing concerns about environmental pollution, the EPA was established on 2 December 1970 to consolidate in one agency a variety of federal government environmental protection activities. Its stated mission is to 'protect human health and the environment' (EPA 2013). The EPA is responsible for enforcing national environmental standards backed up by statute.

Other significant legislation passed during the 'golden age' included the Clean Air Act (1970), the Federal Water Pollution Control Act (1972), the Resource Conservation and Recovery Act (1976) and the Comprehensive Environmental Response, Compensation and Liability Act (1980). In addition a number of other laws concerning wildlife protection, waste management, mining activities and energy policy were passed into law. These have made a significant contribution to the quality of the US environment (EPA 2008; Rosenbaum 2007).

What stood out in this legislation was the high degree of political consensus and strong bipartisan support amongst Democrats and Republicans in the US Congress. A clear example of this was the 1973 Endangered Species Act whose genesis lay in concerns about the threat to populations of the American bald eagle and the bison; two iconic images of the US. The Act was passed unanimously in the Senate and by 345 votes to four in the House of Representatives. It was then signed into law by the then Republican President, Richard Nixon. The 1970s also saw a significant rise in the number of campaigning environmental groups such as Friends of the Earth, Greenpeace and the Sierra Club. This reflected a growing awareness amongst sections of the electorate about the need to take the environmental agenda seriously. This nascent 'green politics' sought to broaden the agenda beyond traditional notions of material happiness; a political approach which has been described as 'post materialism' (Abramson and Ingelhart 1995).

Such bipartisan support for environmental policy 'triggered a profound expansion of government power' (Klyza and Sousa 2008: 1). Such an extension of federal government power was anathema to those public and political forces (principally in the Republican Party and sections of the business community) for whom the American tradition was firmly rooted in individual freedom and limited government. The 'golden age' began to lose its shine as environmental policy became increasingly presented as unwarranted government intervention. It 'became a focal point for political struggle' (Klyza and Sousa 2008: 1) and party politics became increasingly conflicted.

Environmental and climate change policy over two decades: a broad survey

The policy landscape in the US has changed significantly since the 'golden age'. As Klyza and Sousa note, 'Environmental policy making looks very different' (Klyza and Sousa 2008: 1). We now have what can be described as a policy gridlock which is a product of both political and institutional forces. The last 20 years have seen an increasing bitter partisan political divide between the Democrat and Republican parties about the general policy approach the US should adopt in seeking to meet the challenges of the twenty-first century. Specifically, the challenges of climate change and of energy use have brought to light very different perceptions of the role and scope of government.

Adding to this problem is the very nature of the US constitution itself. The US is governed on the principle of the 'separation of powers', with each of the three branches of the Federal government – the Presidency, the Congress and the Supreme Court – having their own defined powers and areas of competence. The 50 states of the union also have their own defined areas of responsibility. Such a system was originally conceived as an attempt to diffuse political power and avert autocracy. But taken together with increasing conflict and political partisanship it has often led to policy gridlock and institutional fragmentation.

Constraints on the policy agenda

In broad policy terms, challenges such as combating climate change and promoting environmental sustainability are complex and multifaceted. They require holistic thinking and effective inter agency working. Governments, not just those in the US, are not generally very good at this. Indeed, as Bailey notes, 'The challenge of engineering significant and coherent change in environmental policy is considerable' (Bailey 2010: 168). This has been clearly illustrated over the last two decades.

This general governance problem has been exacerbated in the US context over the last two decades by a number of key factors. First, the growing partisan divide which has characterised US politics for the last two decades has had a major impact on the policy process. However, it is worth pointing out that in 2008 as Klein argues, 'tackling climate change still had a veneer of bipartisan support' (Klein 2014: 35). In that year, the former Republican speaker of the house, Newt Gingrich, and Democratic congresswoman Nancy Pelosi, 'did a TV spot in which they pledged to join forces and fight climate change together' (Klein 2014: 35). However, over recent years, especially with the election of Barack Obama as President, the issue of climate change has become increasingly ideological, and was in many ways used as a stick with which to beat Obama. Within Congress today, many conservative and Tea Party leaning Republicans (such as Ted Cruz and Michelle Bachman) look upon policies to tackle climate change as nothing but a concerted and underhand attempt to further strengthen the power of the federal government. Viewed through such a lens, international agreements on, for example, the reduction of greenhouse gases, are no more than an attempt to undermine US economic and political sovereignty.

Second, business interests and conservative politicians have long been critical of the overweening power of the federal government and its agencies in respect of environmental laws and regulations and the financial burden that this has imposed on the private sector (Klyza and Sousa 2008: 3). This has had a major impact on the policy debate.

Third, the development of environmental laws from the 'golden age' has created what Bailey has described as a 'green state' made up of lobby groups and vested interests 'that constrain change' (Bailey 2010: 170). This suggests a possible reluctance by some established green groups to adapt to new policy challenges such as climate change, with a focus on protecting their own interests. One might argue that they have become part of a conservative environmental establishment stuck in a mindset of the past, when what is needed is a new mindset to both articulate and respond to the very different challenges that climate change presents.

Fourth, the policy making arena with regard to the environment is crowded with vested interests on various sides of the argument, each pushing their own agenda. This can often result in sub optimal solutions which fail to deal with the problem. In addition, turf warfare between federal government agencies, ideological battles within Congress, Presidential edicts, court rulings and initiatives at the level of individual states have led to a complex policy environment that lacks coherence.

Finally, one of the key constraints to the further development of the environmental agenda over the last two decades in the US is the nature of climate change itself. The challenge it presents is of a different order to what one might term traditional environmental problems, such as air and water pollution. That is not to suggest that the solutions to such problems have been easy. But the challenge of climate change is of a different order. It has been described as a wicked problem, complex and challenging in its nature. As Hulme notes, 'wicked problems frequently emerge from unbounded, complex and imperfectly understood systems. Their solution is beyond the realm of mere technical knowledge and traditional forms of governance'. For Hulme, 'climate change possess all of the attributes of a wicked problem' (Hulme 2014: 119). As such it requires a holistic approach, drawing on politics, society, culture, art, science, human psychology and economics. It requires action at all levels of government, both nationally and internationally. That is why it is such a difficult nut to crack.

Tackling climate change also requires significant sacrifices on the part of the citizens of the US and an almost certain reduction in lifestyles. For a people brought up in a culture of consumerism this is a significant psychological challenge. It is also a difficult sell for politicians. Telling voters they should expect less is not a good way to get elected. It requires a set of brave political decisions and action! For the remainder of this chapter, I will look at the policy record on climate change and the environment of the two most recent US Presidents prior to Barack Obama, namely William (Bill) Jefferson Clinton and George W. Bush.

The impact of environmental and climate change policy in the Clinton era

When the Democratic President Bill Clinton entered the White House in January 1993, there was a general optimism amongst the environmental lobby. The Ronald Reagan and George H. W. Bush Presidencies were the setting for major conflicts over issues ranging from how to tackle acid rain to attempts to weaken the role of the EPA. Vice President Al Gore, however, had a proven track record on environmental causes and was clear in his commitment to take action on climate change. Nonetheless, looking back on the Clinton record, Klyza and Sousa argue that there was 'little in the way of new conversation and environmental laws to mark his eight years as President' (Klyza and Sousa 2013: 97). In large part this is a product of the increasingly partisan and conflicted nature of the US polity which made it more and more difficult to achieve legislative agreement in the US Congress. During Clinton's period in office, in particular the latter part, the political climate became more and more rancorous with open hostility by Congressional Republicans; led by Newt Gingrich, the then speaker of the House of Representatives. Indeed, with a Republican majority, the House voted to impeach the President. In the event, the impeachment process failed in the Senate, but the political atmosphere of the time did much to impede policy action on climate change and the environment. Dunlap and McCright write of 'The anti environmental orientation of the Republican Party' (Dunlap and McCright 2008). The rancorous nature of the debate is well illustrated with two examples pertaining to the EPA. In a speech on the floor of the House of Representatives in 1995, Republican majority whip Tom Delay described the EPA as 'the Gestapo of government' (Bailey 2010: 71). Another example came in September 1995 when the Senate passed an appropriations' bill which would have cut the EPA's funding by some $1.5 billion, severely hampering its ability to enforce clean air regulations. It was passed by 55 votes to 44 but was vetoed by President Clinton in December of that year.

However, despite such opposition, the Clinton Presidency did actively engage in shaping and supporting policies to tackle climate change. On 22 October 1997, Clinton gave a speech to the National Geographic Society outlining his policies on climate ahead of the Kyoto negotiations on the UN Framework Convention on Climate Change (UNFCCC). For Clinton there was

> a clear responsibility and a golden opportunity to conquer one of the most important challenges of the 21st century – the challenge of climate change … with an environmentally sound and economically strong strategy, to achieve meaningful reductions in greenhouse gases in the United States and throughout the industrialized and developing world.
>
> (Royden 2002, p.415)

He went on to talk about the dramatic increase in greenhouse gases since the dawn of the Industrial Revolution arguing that 'the process must be slowed, then stopped, then reduced if we want to continue our economic progress and preserve the quality of life in the United States and throughout our planet. We know what we have to do' (Royden 2002: 416). It was as Royden observes 'an ambitious statement' of intent (Royden 2002: 416). But was there any substantive policy impact? The Clinton administration did indeed set in place a number of programmes which were designed to reduce greenhouse gas emissions. For example, it established the Climate Change Technology Initiative (CCTI). The CCTI was a $6.3 billion, five year package of spending and tax incentives, the aim of which to promote the use of energy efficient technologies in power generation, buildings, motor vehicles and industrial processes. In addition, the Clinton administration played an important role in helping to shape the 1997 UN Kyoto Protocol on climate change. In the Protocol, industrialised nations signed up to a commitment to cut their combined greenhouse gas emissions to 5 per cent below 1990 levels by 2008–2012. Yet, despite the support of the White House, the US Senate failed to ratify the Kyoto Protocol. A subsequent voluntary agreement by the US under the UNFCCC failed to reduce US greenhouse gas emissions by the year 2000 back to 1990 levels.

On a broader policy level, the White House claimed that as a result of the eight years of the Clinton/Gore administration, the US had the cleanest environment in a generation (White House 2000). This is a bold statement indeed. Yet setting aside the political rhetoric, the Clinton Presidency did launch a number of environmental policy initiatives utilising various forms of executive action. One of Clinton's most notable policy proposals was an attempt to protect some 60 million acres of roadless national forest land from future development in the state of Wyoming. As well as being an important conservation measure, it brought with it potentially important co benefits such as limiting greenhouse gas emissions in the protected areas. However, his decision was subject to a number of court actions. In 2009 and 2011, two federal court decisions ruled in favour of the Clinton administration's action. In October 2012, the Supreme Court rejected an appeal by the state of Wyoming, and other interested parties, to overturn the decision of these lower courts. Clinton also used his executive authority to issue new environmental guidelines for hard rock mining and tightened up the regulation with respect to the protection of wetlands. Strict new ambient air quality standards for particulates were introduced, which had the effect of strengthening the Clean Air Act. The Clinton Administration also increased the scope of the Toxic Release Inventory Program (TRI) by doubling the amount of potentially hazardous chemicals that companies should report. The TRI is a publically available database containing information on toxic chemical releases and other waste management activities in the US at the local, state, regional and federal level. The Clinton administration further strengthened aspects of the Clear Air Act through measures such as increased legal restrictions on tailpipe emissions from cars, light vehicles and SUVs, together with a 90 per cent reduction of sulphur in gasoline.

Bill Clinton used his executive authority to set up the Council on Sustainable Development (CSD) to advise him on 'bold, new approaches to achieving our economic, environmental and equity goals (in the US)' (CSD 2017). The CSD produced a document in May 1999 entitled 'Towards a sustainable America'. Setting action on climate change in the context of the broader sustainability agenda, it spoke of the challenge to create 'a life sustaining earth' that supports 'a dignified, peaceful and equitable existence' (CSD 1999). It is a radical and far reaching document. It sets out seven goals, including goals on equity and conservation of nature. Goal 3 on equity aims to 'ensure that all Americans are afforded justice' and have the opportunity to achieve economic, environmental and social well being. Goal 4 on the conservation of nature, talks of the need to 'use, conserve, and protect natural resources … in ways that help ensure the long term social, economic and environmental benefits for ourselves and future generations'. Goal 9 talks of the 'task for the US in taking a leadership role in the development and implementation of global sustainable development policies'. It states that 'The risk of accelerated climate change in the next century has emerged as on one of the most important issues we will face as we seek to achieve our sustainable development goals' (CSD 1999). The policy document sets out a number of strategies for action on climate change arguing that climate change policy 'should be fundamentally linked to any national agenda for economic growth, environmental protection and social justice' (CSD 1999).

Despite the turmoil surrounding Bill Clinton's time in office, his Presidency did leave behind a legacy of action on climate change and environmental protection. Hollern-Harvey has argued that Clinton's policy on the environment was well received by the public during his time in office (Hollern-Harvey 2000: 135). The CSD was dissolved at the end of the Clinton Presidency but as Farley and Smith note, 'It remains the closest the United States has come to a national strategy to date' (Farley and Smith 2014: 96). Indeed, after leaving office Bill Clinton carried on some of its work with the Clinton climate initiative. It is part of the Clinton Foundation and promotes action on climate change by supporting projects in areas such as bio diversity, reforestation and energy efficiency. His successor, George W. Bush, had a much more mixed record (to put it mildly) on climate change action and environmental protection and it is to this that I now turn.

The George W. Bush legacy

Nothing better sums up the contested policy landscape of environmental and climate change policy than the Presidency of George W. Bush. He was elected to the Presidency in November 2000 in highly contested circumstances. Indeed as Grossman *et al.* note, Bush was 'a lightning rod for controversy from the inception of his presidency' (Grossman *et al.* 2009: 1). Climate change policy was no exception. As Bailey notes, 'The election of George W. Bush in 2000 gave a boost to opponents of government action to address climate change' (Bailey 2015: 87). With a background in the oil industry, Bush attracted the opprobrium of the environmental lobby from the very start of his administration. Farley and Smith speak of the overtly negative attitude during the Bush Presidency towards policies on the environment (Farley and Smith 2014: 97). If the American environmental lobby had their own court, then George W. Bush would have been one of the first in the dock. Let us consider the list of charges.

Early appointments to the Bush administration gave a clear sense of the different approach to action on climate change. Those with backgrounds in the energy and automobile industries were to the fore. For example, chief of staff Andrew Card had worked as a lobbyist for General Motors. Commerce secretary, Donald Evans, had close links with an oil and gas company in Denver and the national security advisor, Condoleeza Rice, had served as a director on the board of Chevron. As Bailey notes, this gave 'industry groups a privileged voice in the deliberations and decision making venues of government' (Bailey 2015: 92).

Crucially, less than three months into the Bush Presidency, the White House signalled a major retreat on the climate change policies of the previous Clinton Presidency when it declared that the Bush administration had 'no interest in implementing the Kyoto Protocol' (Bailey 2010: 176). In a 13 March 2001 letter to several members of the US Senate, Bush announced his opposition to the Kyoto Protocol, arguing that it would cost the US around five million jobs if implemented (Aldy 2004). A White House press release in June 2001 confirmed the point (White House 2001). As I noted earlier, the 1997 Kyoto Protocol was a global agreement on climate change in which the signatories agreed to cut their combined greenhouse gases by 5 per cent from 1990 levels through 2008–2012. Bush's Vice President, Dick Cheney (a key player in Administration unlike many previous Vice Presidents), stated his opposition to the principles underpinning the Kyoto Protocol arguing that a CO_2 emissions cap would contribute to energy shortages and was not good energy policy (Draper 2007: 122). Bush did make a promise to regulate the greenhouse gases responsible for global warming during the 2000 Presidential campaign but broke this promise once in office (Tiefer 2004: 280). Lobby group politics played a key role here. For as Tiefer argues, Bush 'brought from Texas and from his Presidential campaign a determination to serve industries seeking to exploit public resources as well as heavy industrial polluters' (Tiefer 2008: 280). Vogel also argues that the unwillingness of the Bush administration to take action on climate change stemmed from the opposition of business (Vogel 2012: 136). I will look the role of lobby groups in more detail in Chapter 3. In his 2010 autobiography, *Decision Points*, Bush attempts to counter the narrative of his unwillingness to tackle climate change. In reference to the summer 2007 G8 summit hosted by the German Chancellor Angela Merkel in which global warming was to be a key agenda item, the President claims he told Merkel that he 'was willing to be constructive on the issue' (Bush 2010: 347). Yet in the same tone he speaks of the 'flawed Kyoto Protocol' and talks of the need to develop 'an international process' focused on clean energy technology to cut greenhouse gas emissions without stifling economic growth (Bush 2010: 347).

This led, as McKay notes, to accusations that the US had become 'an environmental rogue state' (McKay 2009: 405). In similar vein, Tiefer has written of the 'stealth anti environmental policy' of the Bush era (Tiefer 2004: 80). In relation to other broader environmental policies, Klyza and Sousa note how President Bush used executive powers to actively reverse a number of the policy initiatives of the Clinton era (Klyza and Sousa 2013: 98). These included protecting roadless lands in national forests, hard rock mining regulations and regulations around clean drinking water. The Bush administration also sought to weaken the provisions of the Clean Air Act by exempting older utilities and factories from its provisions. In addition, in a clash with the EPA, President Bush rejected the findings of the clean air scientific advisory committee, and sought to lower the benchmark for what constituted clean air. The Bush administration also extended the categories of logging activities in national forests that were not required to meet certain specified environmental standards and lifted an executive ban on off shore drilling that had stood since the last President Bush in 1990. Approval was also given to the Cheney energy task force's pursuit of drilling in the Arctic National Wildlife Refuge. In addition, key positions in the EPA were filled with 'Industry lawyer-lobbyists ... or with people who could be expected to take a laissez faire attitude' (Tiefer 2004: 280). Furthermore, as Bailey notes, President Bush used his executive powers, which resulted in the undermining of 'decades if not a century of progress on the environment' (Bailey 2010: 175), weakening environmental protection legislation in order to favour and benefit mining industries and other others. In addition, as Storey observes, Bush 'eased environmental controls on coal fired power plants, and expanded logging and oil developments on federal government land' (Storey 2010: 195). In the dying days of his second term President Bush, using his executive powers, pushed through a number of potentially environmentally damaging measures. For example, he rescinded a requirement for federal agencies to seek advice from government wildlife experts before opening up new areas for logging or road construction. He also stopped the EPA from carrying out an investigation into the impact of climate change on endangered and protected species (Bailey 2010: 177).

Klyza and Sousa argue that the Bush Presidency 'had something of a free ride in making unpopular conservation and environmental policy' through executive actions (Klyza and Sousa 2013: 99). Indeed, looking back on the George W. Bush era, Carl Pope, executive director of the Sierra Club has argued that 'The effect of the administration's war on terror had been to prevent the people focussing on these issues' (Klyza and Sousa 2013: 99). In addition, policy gridlock in the Congress was a factor in the increasing use of executive orders in the area of environmental policy. As the *New York Times* journalist, Joel Brinkley, wrote 'The administration has often been stymied in its efforts to pass major domestic initiatives in Congress.... So officials have turned to regulatory change' in the shape of executive actions (Brinkley 2004).

A central element of the Bush agenda was actively to undermine the science behind climate change. In 2004 the NASA scientist James Hansen was on record accusing the Bush administration of trying to block the release of data which showed acceleration in global warming. As I have already noted, various lobby groups sought to undermine the science of climate change. Once Bush entered the White House, such efforts began to bear fruit. According to former Democratic US Vice President Al Gore, a long time campaigner on climate change and environmental matters, Bush 'publicly demeaned scientists in his own administration' who wrote official reports that emphasised the dangers facing the US in respect of climate change (Gore 2007: 200). In this regard, Joynt-Kumar notes how scientists working for the NOAA, the US Geological Survey and NASA 'publicly chafed at what the scientists regarded as efforts to interfere with their public descriptions of their own findings' (Joynt-Kumar 2007: 365). In 2006 the Washington Post reported that scientists said they were 'required' to clear all media requests for interviews with administration officials, something which they were not required to do until the summer of 2004 (Eilperin 2006a). Distinguished scientist James Hansen complained that when he posted information to the NASA web site showing that 2005 might be the warmest year since records began, he was ordered to take it down 'because he had not had it screened by the administration in advance' (Eilperin 2006b). There are other examples. A 2001 report published by the highly respected National Academy of Sciences (and requested by the White House) confirmed the consensus on human made climate change with a high degree of scientific certainty, but with the caveat that in the scientific world one can never be 100 per cent certain (National Academy of Sciences 2001). However, as Mooney notes the White House seized on this report and overplayed the claim that uncertainty remains as to the causes of climate change (Mooney 2005: 214).

More generally, as Mazmanian and Kraft argue, under the Bush Presidency there were 'a range of legislative, administrative and judicial assaults on environmental policy' (Mazmanian and Kraft 2009: 20). Evidence suggests that the Bush administration interfered in the terminology that was used in official government scientific reports. Equally alarming is the claim that several sections of EPA reports were censored after interventions by the White House (Mooney 2006).

So how can we sum up the record of George W. Bush? At the end of his Presidency, the White House published a document setting out what it claimed to be his achievements whilst in office. 'Throughout his Administration', claimed the document, 'President Bush made protecting the environment for future generations his top priority' (White House 2009: 27). Indeed Bush does deserve some credit for designating some 140,000 square miles of the north western Hawaiian islands as a marine national monument. The document also talks of how the Bush administration supported the development of clean energy technologies and worked to improve air quality. There is also reference to the commitment of the Bush administration to protect the nation's forests and to preserve its wetlands. The document also states that President Bush sought to 'confront climate change through a rational and balanced approach' (White House 2009: 28). Taken as read, this sounds like a good record, but a lot of the evidence which is presented above points in the opposite direction.

Commenting on the Bush record, the National Resources Defence Council (NRDC) has argued that his administration led the most thorough and destructive campaign against America's environmental safeguards in the last 40 years (NRDC 2005). Shepard and Fears talk of the failure of the Bush administration 'to see the big picture in reference to global environmental change' (Shepard and Fears 2011). They point to the US withdrawal from the Kyoto Protocol as a clear example of this. In a real sense, Bush led the US away from effectively responding to the challenges of climate change. In a damning critique, Josh Dorner of the Sierra Club argues that the Bush administration undid 'decades, if not a century, of progress on the environment' (Goldenburg 2009). Jorner is of the view that Bush 'undermined the rule of law, undermined science ... and rendered government agencies unable to do their most basic functions, even if they wanted to' (Goldenburg 2009).

The Bush administration's antipathy to climate change and environmental policy is neatly summarised in the phrase *drill baby drill.* It was first used as a campaign slogan at the 2008 Republican Convention by the former Maryland

Lieutenant Governor, Michael Steele. The slogan expressed support for increased drilling for oil. Presidential opposition to environmental and climate change treaties continued to the end of the Bush Presidency. But in 2007 the screening of the documentary *An Inconvenient Truth* played its role in pushing environmental and climate change politics further up the policy agenda. It was an agenda that was to gain traction in the subsequent Obama Presidency, albeit with many twists on turns on the way. This is the subject of the next chapter.

Conclusion

The 1992 UN Conference on Environment and Development (more commonly known as the Rio Summit), the accompanying climate change convention and the subsequent Kyoto Protocol on climate change were, or so it appeared at the time, significant steps on the road to a more sustainable world with global agreements to tackle climate change and reduce greenhouse emissions. Paradoxically, in the US domestic context, from the mid 1990s through to 2008, it was one of retrenchment on environmental and sustainability policy and legislation on climate change, particularly so within sections of the Republican dominated Congress. The 'golden age' of 1964 to 1980, underpinned by a cross party consensus, gave way to a conflicted and partisan polity in which environmental policies came under sustained attack, principally from conservative Republicans and supporters of the Tea Party. It was an attack that took on an increased intensity under the Presidency of George W. Bush, with the undermining of the science around climate change and the promotion of policy initiatives which supported mining and drilling interests and weakened a number of the environmental gains of the 'golden age'. This attack on the science of climate change was also supported by powerful lobby groups. Former US Vice President Al Gore has spoken about Exxon Mobil's 'brazen efforts to try to manage public perceptions of the reality and seriousness of the climate crisis' (Gore 2007: 201). The UK-based Royal Society also formally requested that Exxon Mobil stop putting out 'very misleading' and 'inaccurate' information into the public domain (Gore 2007: 201).

In January 2009, Barack Obama was inaugurated as the 44th President of the US. He came into office determined to break with the record of his predecessor. He spoke of the threat that climate change presented to the US and the world and stressed the need for action by the US and the global community. The analysis of the Obama Presidency is the subject of the next chapter.

References

Abramson, P. and Ingelhart, R. (1995) *Value Change in Global Perspective*, Michigan, MI: University of Michigan Press.

Aldy, J. (2004) 'Saving the Planet Cost-Effectively: The Role of Economic Analysis in Climate Change Mitigation Policy' in Lutter, R. and Shogren, J. (eds), *Painting the White House Green: Rationalizing Environmental Policy Inside the Executive Office of the President*, Washington, DC: Resources for the Future.

Atkinson, H. (2015) 'Climate Change and Environmental Policy in the US: Lessons in Political Action' in Atkinson, H. and Wade, R. (eds), *The Challenge of Sustainability: Linking Politics, Education and Learning*, Bristol: Policy Press.

Bailey, C. (2010) 'Environmental Politics and Policy' in Peele, G., Bailey, C., Cain, B. and Guy Peters, B. (eds), *Developments in American Politics*, Basingstoke: Palgrave Macmillan.

Bailey, C. (2015) *US Climate Change Policy*, Farnham: Ashgate Publishing Limited.

Bierman, F. (2014) *Earth System Governance: World Politics in the Anthropocene*, Cambridge, MA: MIT Press.

Bonds, E. (2016) 'Losing the Arctic: The US Corporate Community, the National Security State, and Climate Change' *Environmental Sociology*, vol. 2, issue 1, pp. 5–17.

Brinkley, J. (2004) 'Out of spotlight Bush overhauls US regulations' *New York Times*, 14 August.

Bush, W. G. (2010) *Decision Points*, New York: Crown Publishers.

Draper, R. (2007) *Dead Certainty: The Presidency of George W. Bush*, New York: Free Press.

Dunlap, R. and McCright, M. (2008) 'A Widening Gap: Republican and Democrat Views on Climate Change' *Environment: Science and Policy for Sustainable Development*, vol. 50, issue 5, pp. 26–35.

Eilperin, J. (2006a) 'Climate Change Researchers Feeling the Heat from the White House' *Washington Post*, 6 April.

Eilperin, J. (2006b) 'Debate on Climate Shifts to Issue of Irreparable Change: Some Experts see Tipping Point When it is Too Late' *Washington Post*, 29 January.

Farley, M. and Smith, Z. (2014) *Sustainability: If it's Everything, is it Nothing?*, London: Routledge.

Flavin, C. and Engelman, R. (2009) 'The Perfect Storm' in Worldwatch Institute, *State of the World 2009: Confronting Climate Change*, London: Earthscan.

Friedrich, J. and Damasson, T. (2014) *The History of Carbon Dioxide Emissions*, Washington, DC: World Resources Institute.

Gertner, J. (2008) 'The Future is Drying Up' *New York Times*, 21 October.

Gore, A. (2008) *The Assault on Reason*, New York: Penguin Books.

Grossman, M., Matthews, R. and Cunion, W. (2009) 'Introduction' in Grossman, E. and Matthews, R. (eds), *Perspectives on the Legacy of George W. Bush*, Newcastle upon Tyne: Cambridge Scholars Publishing.

Grover, W. and Peschek, J. (2014) *The Unsustainable Presidency: Clinton, Bush, Obama and Beyond*, New York: Palgrave Macmillan.

Hollern-Harvey, D. (2000) 'The Public View of Clinton' in Schier, S. (ed.), *The Post Modern Presidency: Bill Clinton's Legacy in US Politics*, Pittsburgh, PA: University of Pittsburgh Press.

Hulme, M. (2014) *Can Science Fix Climate Change?*, Cambridge: Polity Press.

Joynt-Kumar, M. (2007) 'Managing the News: The Bush Communications Operation' in Edwards, G. and King, D. (eds), *The Polarised Presidency of George W. Bush*, Oxford: Oxford University Press.

Klein, N. (2014) *This Changes Everything: Capitalism vs. the Climate*, New York: Simon & Schuster.

Klyza, C. and Sousa, D. (2008) *American Environmental Policy, 1990–2006: Beyond Gridlock*, Cambridge, MA: MIT Press.

Klyza, C. and Sousa, D. (2013) *American Environmental Policy Beyond Gridlock*, Cambridge, MA: MIT Press.

IPCC (2013) *Climate Change 2013 Synthesis Report; Evaluation of Working Groups I, II and III to the Fifth Assessment Report*, Cambridge: Cambridge University Press.

McKay, D. (2009) *American Politics and Society*, Oxford: Wiley-Blackwell.

McKie, R. (2016) 'Scientists Warn World Will Miss Key Climate Targets' *Observer*, 7 August.

Mazmanian, D. and Kraft, M. (2009) *Towards Sustainable Communities: Transition and Transformation in Environmental Policy*, Cambridge, MA: MIT Press.

Mooney, C. (2006) *The Republican War on Science*, New York: Basic Books.

NRDC (2005) *Rewriting the Rules; the Bush Administration's First-Term Environment Record*, New York: NRDC.

Reich, R. (2008) *Supercapitalism: The Battle for Democracy in an Age of Big Business*, Sydney: Allen & Unwin.

Revkin, A. (2008) 'New Climate Report Foresees Big Changes' *New York Times*, 28 May.

Rosenbaum, W. (2007) *Environmental Politics and Policy*, Washington, DC: CQ Press.

Royden, A. (2002) 'United States Climate Policy under President Clinton: A Look Back' *Golden Gate University Law Review*, vol. 32, issue 4, pp. 415–478.

Stern, N. (2006) *Stern Review on Economics of Climate Change*, London: HMSO.

Storey, W. (2010) *US Government and Politics*, Edinburgh: Edinburgh University Press.

Tiefer, C. (2004) *Veering Right: How the Bush Administration Subverted the Law for Conservative Causes*, Berkeley: University of California Press.

Vogel, D. (2012) *Regulating Health, Safety and Environmental Risk in Europe and the United States*, Princeton, NJ: Princeton University Press.

Worldwatch Institute (2009) *State of the World 2009: Confronting Climate Change*, London: Earthscan.

Internet sources

Carrington, D (2017) 'Arctic Ice Falls to Record Winter Low After Polar Heat Waves' *Guardian*, 22 March 2017. Available at: www.theguardian.com/environment/2017/mar/22/arctic-ice-falls-record-winter-low-polar-heatwaves?CMP=Share_AndroidApp_Gmail, accessed 23/03/17.

CSD (1999) 'Towards a Sustainable America'. Available at: https://clinton2.nara.gov/PCSD/Publications/tsa.pdf, accessed 15/10/16.

CSD (2017) 'President Clinton's Council on Social Democracy'. Available at: www.numbersusa.org/pages/president-clintons-council-sustainable-development, accessed 02/02/17.

EPA (2013) Available at: www.epa.gov/aboutepa/whatwedo/html, accessed 23/04/13.

Goldenburg, S. (2009) 'The Worst of Times: Bush's Environmental Legacy Examined' *Guardian*, 16 January 2009. Available at: www.theguardian.com/politics/2009/jan/16/greenpolitics-georgebush, accessed 02/02/13.

IEA (2011) 'The World is Locking Itself into an Unstoppable Energy Future Which Would Have Far Reaching Consequences, IEA Warns in Latest World Energy Outlook' press release, 9 November 2011. Available at www.iea.org/newsroomand events/pressreleases/2011/November/name,20318,en.html, accessed 30/11/11.

National Academy of Sciences (2001) 'Joint Science Academies Statement: Global Response to Climate Change'. Available at: http://nationalacademies.org/onpi/06072005.pdf, accessed 07/09/16.

NCADAC (2013) 'US National Climate Assessment Report'. Available at: www.ncada.globalchange.gov.us, accessed 02/02/14.

NIC (2017) 'Global Trends, Paradox of Progress Links'. Available at: www.dni.gov/files/images/globalTrends/documents/GT-Main-Report.pdf, accessed 06/05/17.

NOAA (2013) 'NCDC Announces Warmest Year on Record Contiguous US'. Available at: www.ncdc.noaa.gov/news/ncdc-announces-warmest-year-record-contiguous-us, accessed 04/05/17.

NOAA (2017a) '2016 was 2nd Warmest Year on Record for US'. Available at: www.noaa.gov/news/2016-was-2nd-warmest-year-on-record-for-us, accessed 04/05/17.

NOAA (2017b) '2016 Marks Three Consecutive Years of Record Warmth for the Globe'. Available at: www.noaa.gov/stories/2016-marks-three-consecutive-years-of-record-warmth-for-globe, accessed 04/05/17.

PBL Netherlands (2015) 'Trends in Global CO_2 Emissions 2015 Report'. Available at: http://edgar.jrc.ec.europa.eu/news_docs/jrc-2015-trends-in-global-co2-emissions-2015-report-98184.pdf, accessed 03/12/16.

Romm, J. (2017) 'Global Warming Made Every State a Red State in 2016'. Available at: https://thinkprogress.org/global-warming-made-every-state-a-red-state-in-2016-53e699ca9f0f, accessed 14/03/17.

Shepard, D. and Fears, N. (2011) 'US Environmental Policy and Leadership 2011'. Available at: www.brighthub.com/environment/science-environmental/articles/39623.aspx, accessed 02/02/17.

White House (2000) 'The Clinton Presidency: Protecting our Environment and Public Health'. Available at: https://clinton5.nara.gov/WH/Accomplishments/eightyears-08.html, accessed 23/03/17.

White House (2001) Available at: www.whitehouse.gov/news/releases/2001/03/20010314.html, accessed 02/10/15.

White House (2009) 'The Administration of President George W. Bush, 2001 to 2009'. Available at: https://georgewbush-whitehouse.archives.gov/infocus/bushrecord/documents/legacybooklet.pdf, accessed 02/09/16.

2 The Obama Presidency

Introduction

The election of Barack Obama as the 44th President of the US in 2008 seemed to presage a more constructive engagement by the US with the global community on the climate change agenda. In addition, it seemed to point to a new activism at the federal level with regard to climate change and sustainability. There were three interrelated aspects of Obama's policy position on climate change. First, he attempted to restore the image of science and the facts behind climate, areas that were undermined during the Presidency of George W. Bush. Second, there was an increasing emphasis on the damage climate change could bring to the economy if substantive action was not taken. Third, political action on climate change can be seen in the broader context of Obama's mantra to 'bring change' to America.

Action on climate change in Obama's first term

Obama signalled early on in his administration his determination to give a high priority to action on climate change. In a speech on employment, climate change and energy independence on 26 January 2009, Obama argued that

> Year after year, decade after decade, we've chosen delay over decisive action. Rigid ideology has overruled sound science. Special interests have overshadowed common sense.... For the sake of our security, our economy and our planet, we must have the courage and the commitment to change.
>
> (quoted in Bailey 2010: 168)

He received the backing of all the major environmental lobby groups with opinion polls consistently showing that voters thought he had the better policies on the environment and energy. As Dunlap *et al.* note, the election of Obama 'created great optimism among supporters of many progressive causes, including environmental protection and action on climate change' (Dunlap *et al.* 2016). Indeed, his Vice President Al Gore had put the issue of climate change firmly on the political agenda with the release in 2006 of the documentary film *An Inconvenient Truth*, which set out in clear terms what the consequences would be for people and planet if immediate action was not taken to combat climate change. A book followed in 2007. In October 2007, Al Gore was award the Nobel Peace Prize for his work highlighting the threat posed by climate change.

Light argues that 'Presidents have a degree of choice in selecting agenda issues' (Light 1999: 63). Whilst this may be true, Obama faced a number of considerable hurdles in his attempt to push action on climate change up the policy agenda. There were a number of significant and substantive policy initiatives with regard to tackling climate change but the reality of partisan politics and the resultant legislative gridlock in the US Congress served to weaken the Presidential agenda at times. It also has to be said that Obama's somewhat cautious approach to policy making did at times act as something of an obstacle. But the chapter will argue that the journey of travel with regard to tackling climate change was heading in the right direction, however stuttering it was at times.

Obama took over the reins of office in January 2009. He had been elected in the most difficult of circumstances, with the global financial crisis and credit crunch rocking the very foundations of both the US and the global economy. As Bailey notes, Obama 'signalled his intention to break decisively with the policies of his predecessor and give environmental issues, particularly energy reform and climate change, a high priority in his administration' (Bailey 2010: 167). Indeed, 'within his first week in office Obama had signalled a marked departure from Bush era environmental politics' (Bomberg and Super 2009). In his inaugural address on 20 January 2009, he talked of the environmental threats that excessive American energy use posed to the planet. He made reference to the 'spectre of a warming planet'. He also spoke of his ambition to 'harness the sun and the winds and the soil to fuel our cars and run our factories'. As Klyza and Sousa note, when Obama was elected to the Presidency, 'executive politics tilted towards environmentalists' (Klyza and Sousa 2013: 297). This was reflected in some of his early appointments. These included the Noble physics laureate Steven Chu as Energy Secretary while Lisa Jackson, a former commissioner of the New Jersey Department of Environmental Protection, became the new head of the Environmental Protection Agency (EPA). In addition Carol Browner, a former administrator in the EPA in the Clinton administration, was given the task of coordinating action on climate change across various government agencies. Van Jones, a leading advocate in the environmental movement was appointed to oversee the development of the so-called green economy and green jobs, a so-called green czar.

Addressing the UN General Assembly in September 2009, Obama spoke of the necessity of those wealthy nations, including the US, who had 'done so much damage to the world in the 20th century' to lead in the fight on climate change. He also spoke of the need 'to help the poorest nations both to adapt to the problems that climate change has already wrought and help travel on a path of clean development'. Obama acknowledged the major challenge that this presented, not least in 'the midst of a global recession' and that it would 'be tempting to sit back and wait for others to move first. But,' he argued, 'we cannot make this journey unless we all move forward together.'

In the first two years of his Presidency, Obama did push for legislative action in Congress to tackle climate change, in particular action to reduce greenhouse gases. With Democratic majorities in the House and the Senate, 'the moment seemed ripe for a new law controlling carbon emissions' (Klyza and Sousa 2013: 298). Indeed Obama did win support in Congress for significant investment in clean and renewable energy technologies as part of the emergency stimulus bill designed to mitigate the impact of the global financial crisis on the American economy. Yet Obama faced criticism from the environmental lobby for his failure to get a Democratic Congress to pass a climate bill. In the first two years of the Obama Presidency, the Democrats had a majority in both the Senate and the House of Representatives. Whilst such criticism has validity, Presidential attempts to push action on climate change have to be seen in a broader context. In his first term Obama spent much of his political capital on health care reforms and on stimulus programmes in response to the economic recession. The economic crisis that had erupted during the 2008 Presidential campaign also served to focus concerns about job prospects and energy costs. This led to large scale opposition amongst many Republicans and from some Democrats representing areas where oil and coal made an important contribution to the local economy.

But crucially the reality of partisan politics cast a shadow during Obama's first term, with many in the Republican Party – Tea Party supporters in particular – determined at every turn to thwart much of Obama's policy agenda. Action on climate change was no exception. There was an increasingly vociferous climate change denial lobby within the Republican Party. For example, when Obama first came to office in January 2009, he favoured a cap and trade approach to tackling climate change. It was a market based system with permits sold to greenhouse gas emitters. The less a company polluted, the less it would have to pay for permits. It was a policy which, according to Sussman and Daynes, Obama 'hoped would be a mechanism to attract support from both sides of the aisle' (Sussman and Daynes 2013: 59). In other words Obama saw it as a way to build a consensus (a key aspect of Obama's approach to politics) between Democrats and Republicans in Congress to reduce greenhouse gas emissions. But Obama's plans faced strong opposition very early on into his administration and failed to gain sufficient support in the Senate. This was largely the story of Obama's first term in office.

Take for the example the 2009 American Clean Energy and Security Bill. Although there was no reference to climate change in the title of the bill, one of its key provisions was to set a price on carbon. It also laid out targets for cuts in greenhouse gas emissions in two stages. Stage one involved a 17 per cent cut in emission from 2005 levels by 2020. Stage two involved an 80 per cent cut from 2005 levels by 2050. The bill was passed in the House of Representatives by 219 to 212 votes, as there was a Democratic majority at this time. It was described by the Environmental Defense Action Fund (an environmental advocacy group) as 'the most important environmental measure in the last 20 years' (Sussman and Daynes 2013: 61). The bill had the strong backing of President Obama. However, it soon ran into legislative trouble. The bill failed to pass in the Senate despite the support of the influential Senate Environmental Committee. Indeed, as Cohen and Miller note, the US Congress had reached a virtual impasse on environmental and related legislation by the end of Obama's first term of office (Cohen and Miller 2012: 42).

Furthermore, during Obama's first term in office there were a number of legislative attempts to actually roll back action on climate change, again a product of the increasingly rancorous political partisanship and the influence of big oil and the climate change denial lobby that I referred to in Chapter 1. One such example was The Energy Tax Prevention Action which was introduced into Congress on 3 March 2011. As Sussman and Daynes note, it 'was one of the more notable bills directed at the EPA' (Sussman and Daynes 2013: 58). Its central purpose was to effectively prevent the EPA's authority to regulate greenhouse gases under the provisions of the Clean Air Act. This despite a 2007 Supreme Court decision (Massachusetts v EPA) which ruled that the EPA had the legal authority under the Clean Air Act to regulate greenhouse gas emissions. I shall return to the role of the courts and climate change policy in Chapter 3.

Such a proposal, to weaken the role of the EPA, found favour with many Congressional Republicans, especially those linked to the Tea Party. But it is also a position by a number of Democratic legislators representing so-called oil and coal states. One such example is that of Jay Rockefeller, Democratic senator for the state of West Virginia, from 1985 to 2015, a state rich in coal reserves which played a key role in the local economy. In January 2011, he introduced a bill into the Senate which sought to impose a two year delay to plans by the EPA to regulate greenhouse gases under the provisions of the Clean Air Act. There are other examples of attempts to undermine environmental and climate change legislation in Obama's first Presidential term. Henry Waxman, a Democratic congressman from 1975 to 2015 and a senior member of the Energy and Commerce Committee published a database of over 300 anti-environmental votes by members of the 112th Congress, which ran from January 2011 to January 2013. Waxman saved his particular opprobrium for the House of Representatives (the second chamber of Congress). 'This is the most anti-environmental House in history', he argued (Sussman and Daynes 2013: 59).

As a consequence of this legislative gridlock and policy stasis, substantive action on climate change at the federal level in the US was more and more centred in the realm of executive politics with the likes of Presidential directives and administrative actions. In using executive actions, Obama may have demonstrated his constitutional authority, but his use of such executive actions also reflected a political weakness as regards action on climate change and the environment. Obama issued a number of Presidential orders shortly after taking office which reversed a number of the environmental policy initiatives of the Bush era. These included the introduction of stricter codes for clean air, action on greenhouse gas emissions, restrictions on off shore drilling and drilling on public lands near national parks, environmental protection for endangered species, and higher fuel efficiency standards. In late 2009, the Obama administration also granted a so-called waiver to the state of California, thus allowing California to set its own air quality standards under the terms of the 1970 Clean Air Act. There were a number of other initiatives. In November 2011, for example, Obama proposed fuel efficiency standards for smaller vehicles of 54.5 miles per gallon by 2025. In March 2012, the EPA set out higher standards for CO_2 emissions from new power plants.

There were also a number of significant policy announcements on climate change and the environment in the first term of the Obama Presidency. For example, on 26 January 2009 the administration published its 'New Energy for America' plan (also known as the Biden Obama energy plan). The policy document talked of the energy challenges facing the US and how such challenges had gone unaddressed for too long. It spoke of the US addiction to foreign oil and how this had undermined both the environment and national security. The policy document set a target of an 80 per cent reduction in US greenhouse gas emissions by 2050 from 1990 levels with an interim target of 30 per cent reduction by 2025. It outlined plans to invest 150 billion dollars in the green economy, generating five million jobs over ten years (White House 2009). There was also reference to a plan to put one million plug-in hybrid cars on the road by 2020, cars that would be built in America. In addition, it set a target of 25 per cent of electricity usage to be provided by renewables by the year 2025. The stated aim of the policy document was to make the US a world leader in tackling climate change. Key to this was the need to implement a market based cap and trade scheme. However, as I noted above, this specific proposal failed to get approval within Congress.

In 2011, the White House issued the policy document *Blue Print for a Secure Energy Future.* It argued that 'Leading the world in clean energy is critical to strengthening the American economy and winning the future' (White House 2011: 4). It talks about harnessing the US clean energy potential so that 80 per cent of electricity will come from clean energy sources by 2035. There is also reference to the need to produce more fuel efficient cars and trucks and to make buildings more energy efficient. In all of these objectives the federal government needs to lead by example.

Action on climate change during Obama's second term

In November 2012, the President was re-elected to serve a second term. During the preceding election campaign Obama took part in three televised debates with his Republican rival, Mitt Romney. The issue of climate change did not feature in any of the debates. Despite this inauspicious start, Obama, faced with continuing opposition from Republicans in a divided Congress, signalled his intention to push ahead with action on climate change. Indeed during the election campaign the then Mayor of New York, Michael Bloomberg (former Republican turned independent), gave his support to Obama 'citing Republican challenger Mitt Romney's failure to back climate change measures' (MacAskill 2012).

Writing in 2013, Sussman and Daynes speak of President Obama's 'desire that legislation (on climate change) come from Congress if at all possible' (Sussman and Daynes 2013: 60). Yet he made clear his determination to use his executive powers to put action on climate change higher up the policy agenda. He seemed determined to shake off what many perceived to be a failure to take substantive action on climate change in his first term. He brought into his administration individuals with a strong commitment to action on climate change, such as John Kerry who was appointed Secretary of State. On 16 February 2014 on an official visit to Indonesia, Kerry gave one of the strongest speeches on climate change by a member of the US administration. He argued that those who denied the reality of climate change were 'simply burying their heads in the sand'. Kerry went on to argue that climate change was 'the greatest challenge of our generation' and that it 'can now be considered another weapon of mass destruction' (CNN 2014).

There were a number of early indicators that tackling climate change and promoting sustainability were beginning to move up the Obama policy agenda once more. In his State of the Union address to Congress on 12 February 2013, Obama made a detailed case for addressing climate change, arguing that:

> We can choose to believe that super storm Sandy and the most severe drought in decades, and the worst wildfires some states have ever seen were all just a freak coincidence. Or we can choose to believe in the overwhelming judgement of science … and act before it is too late.

Obama called on Congress to draft appropriate legislation but made it clear he was prepared to craft executive actions if such legislation was not forthcoming.

In June 2013, the President's climate action plan was published. It talks of the moral obligation to future generations to leave them a planet that is not polluted. In the document, Obama reiterated the commitment of the White House to reduce US greenhouse gases by 17 per cent below 2005 levels by 2020. It also said that his administration would work 'intensively to forge global responses to climate change' through a series of international agreements and negotiations (White House 2013: 17). There is reference to the key leadership role that the US can play the development of renewable energy by supporting innovation and invention in the new technology sector. The document sets out proposals for cutting carbon pollution from power plants which account of 32 per cent of all US greenhouse gas emissions. There are also targets set for improving the fuel efficiency of heavy duty vehicles and for curbing emissions of hydro fluorocarbons (HFCs), a significant greenhouse gas. On 22 June, in a speech at Georgetown University, Obama outlined in clear terms his strategy for tackling climate change. In effect he served notice that he would be making extensive use of executive orders. This included his decision to bypass Congress and issue an executive memo to the EPA calling for rules to curb greenhouse gas emissions from power plants. The speech was, as Grunwald argues, 'a notification of actions taken and actions to come, actions that don't require help from Congress'

(Grunwald 2013). In October 2013, for example, Obama issued an executive order setting up a 'Task Force on Climate Preparedness and Resilience'.

In the face of continuing opposition from Republicans in Congress he told the assembled students at Georgetown University that he refused 'to condemn your generation and future generations to a planet that's beyond fixing' (Obama 2013). The President also announced major developments to support and promote renewable energy sources (White House 2013).

Keystone XL Pipeline

Originally proposed in 2008, the Keystone XL pipeline was to carry oil from the tar sands of Alberta Canada to Texas. It was to prove a major challenge for the Obama administration and was seen very much as a litmus test of his environmental credentials. It became something of a political lightning rod with vocal opposition from a number of high profile environmental groups including the Sierra Club, Friends of the Earth and the NRDC, co-ordinated by the campaigning organisation 350.org. Oil extracted from tar sands is 20 per cent more carbon intensive than oil produced from traditional drilling methods, and if used as a fuel would result in an increase in greenhouse gases with a concomitant impact on climate change. A decision to approve or reject the pipeline fell within the remit of the President. What would he do? For Klyza and Sousa, 'Obama was stuck' (Klyza and Sousa 2013: 302). They argue that Obama wanted to approve the pipeline as it would bring economic and energy benefits. Whilst such a view is open to challenge, Obama did face a number of pressures. Obama was confronted by a powerful oil lobby with strong support in the Republican Party. In addition a number of labour unions, traditional supporters of the Democrats, argued for the job benefits of the pipeline. The environmental lobby put intense pressure on Obama to block the pipeline. It became a touchstone issue with regard to Obama's commitment to the environmental and climate change agenda. Mass civil disobedience at the White House in August 2011 led to some one thousand arrests. In November 2011 the President announced a delay in the decision until after the 2012 Presidential and Congressional elections. Indeed, during the 2012 campaign, Republican Presidential candidate Mitt Romney promised to approve the pipeline on his first day in office. However, in early November 2015, seven years of uncertainty was brought to an end as Obama made the decision to block Keystone XL. However, the story does not end there. In March 2017, newly elected President Trump signed an executive order giving the go ahead to the pipeline. However this decision has met with strong opposition from the environmental lobby and will be the subject of action in the courts for some time to come. Indeed on the 30 March 2017, several environmental groups filed lawsuits challenging the Trump executive order. In one legal submission, environmental groups including the Centre for Biological Diversity, the Sierra Club and the Northern Plain Research Council argued that the decision to grant a permit to build the pipeline violated the National Environmental Policy Act (Reuters 2017).

Clean power plan

The CPP is arguably one of the most important environmental policy initiatives in the US since the so-called golden age. It was first proposed by the EPA in June 2014. The final version launched by Barack Obama with EPA administrator, Gina McCarthy on 3 August 2015. In an official press release the White House spoke of the CPP as an example of continued US leadership on climate change (White House 2015). In similar vein, the EPA argue that the plan 'is an historic and important step in reducing carbon pollution from power plants that takes action on climate change' which 'shows the world that the United States is committed to leading global efforts to address climate change' (EPA 2016a). For the White House there was a 'moral obligation to leave our children a planet that's not polluted or damaged'. The environmental and health risks of climate change are so stark that 'taking action now was critical' (White House 2015).

The plan runs to some 645 pages and was designed as a key part of the US commitment to the United Nations conference on climate change (UNFCCC) in Paris in December, the so-called COP21. It is also linked to the growing accord between the US and China in the area of climate change during the Obama administration. Back in February 2014, Secretary of State John Kerry announced in the Chinese capital Beijing that the two nations would cooperate on efforts to 'move the climate process forward' in advance of the COP21 in Paris in December 2015 (CNN 2014). Despite a rather tetchy relationship in aspects of foreign such as the dispute over the South China Sea, action on climate change has been a positive aspect of US/China relations. I will focus on this in more depth in Chapter 6 when I consider the US role in global climate change politics.

The CPP is a complex and detailed document but we can draw out some of its key themes. It creates the first ever set of national standards in the US that address carbon emissions from the nation's power plants. Central to the plan is a commitment to reduce carbon emissions by 32 per cent below 2005 levels from power plants by 2030. It involves a partnership between the EPA and the 50 States of the Union with the EPA setting targets for each individual state with implementation on the ground based on local conditions. However ambitious the plan might appear, some words of caution are needed. Whilst the CPP is a significant policy development it still sees a major role for fossil fuels in the energy mix of the US.

The EPA makes the point that fossils fuels will continue to be a critical component of America's energy future. But it then goes on to argue that 'the transition to clean energy is happening faster than anticipated' (EPA 2016a). Indeed 20 per cent of coal fired power plants are expected to close by 2020. Whether newly elected Trump's public support for the coal industry will change this scenario remains a matter for conjecture. The announcement of the CPP was subjected to a great deal of lobbying with environmental groups to the fore. As a consequence Deyette argues that the final draft was significantly strengthened with a bigger role given to the role that renewables can play in carbon reduction. For Deyette wind, solar and other forms of clean energy are well positioned to help states meet their carbon emissions targets (Deyette 2015). Indeed the CPP introduces a number of incentives for individual states to develop more renewable energy.[1] Rick Umoff, an adviser at the US Solar Energy Industries Association, was confident that the new rules would shift investment towards solar. Indeed the aim is to increase renewable energy by 30 per cent by 2030 (White House 2015). The Plan aims to cut sulphur dioxide levels in the atmosphere by 90 per cent and nitrogen emissions by 72 per cent by 2030, with 2005 as the baseline.

The increased emphasis on renewable energy by the EPA and the Obama administration has caused considerable concern, disquiet and no little anger in the hydraulic fracking industry.[2] For Marty Durbin, President of America's Natural Gas Alliance, the reported shift in favour of renewable energy 'is perpetuating a false choice between renewables and natural gas'. He goes on to argue that 'An accelerating move to natural gas is critical to keeping the lights on' (Durbin 2015). Indeed the debate between the exigencies of tackling climate change on the one hand and energy security on the other hand is a complex one. We shall return to this in Chapter 4.

The CPP has faced a number of legal challenges since it was launched. On the 23 October 2015, 24 state governments, electric power producers, trade associations and coal companies filed suits against the EPA seeking a stay (delay) of the CPP. At the same time, the NRDC, an environmental advocacy group, joined 18 state governments, 12 power companies, a number of clean energy associations and a variety of environmental groups to oppose the stay. The legal action over the CPP is just one example of the contested nature of climate change and environmental policy in the US with pressure groups and lobbyists seeking to exert their influence. I will return to this in more detail in Chapter 3.

So who was behind the initial legal action to block the CPP? Auel notes that while some power companies support the CPP and others remain neutral in the law suit, the majority of opposition is from power producers and the trade associations (Auel 2016). Doniger makes a similar argument (Doniger 2016). Key power companies involved in the litigation include the Southern Company (a large integrated electricity utility), NRG Energy inc. and Energy Future Holdings Corp. Other energy companies were involved in the litigation through membership of various trade associations. These included the American Coalition for Clean Coal Electricity (ACCCE), the American Public Power Association and the National Rural Electric Co-operative Association (NRECA) (Auel 2016).

There were also 24 state governments involved in the Supreme Court action. They included Florida, Ohio and Texas. I will look at the interplay between state governments and lobbyists around climate change and environmental policy in more detail in Chapter 3 together with an analysis of the state of public opinion. However, an analysis of public opinion on the CPP does throw up some interesting findings. A 2016 survey found that 66 per cent of registered voters in the states which were party to the Supreme Court legal action were actually in favour of the plan. The survey was conducted by the Program for Public Consultation (PPC) at the University of Maryland. 'Clearly the forces driving this lawsuit' argued PPC director Steven Kull 'are not arising from public resistance to the CPP' (Robinson 2016). Of note is the fact that the survey found that, among respondents who either work in the coal industry or have family members that do, support for the CPP still stood at 62 per cent. However, attitudes to the CPP do reveal a partisan split. Support for the CPP is at highest amongst Democratic voters with a figure of 89 per cent. Support amongst Republican voters lags behind on 47 per cent. We also find this partisan divide across the climate change and environmental policy agenda more generally. I will return to this issue in Chapter 3.

On 9 February 2016, the Supreme Court by a majority of five to four stayed the implementation of the CPP pending judicial review. To further complicate the picture, newly elected President Trump signed an executive order which sought to rescind the CPP. However, this executive order has already been the subject of legal challenges by the environmental lobby and a number of cities and states. The legal wrangling is certain to continue for some time to come. Nonetheless, despite such legal toing and froing, the fact remains that on the ground a number of energy companies will continue to invest in clean technology). For example, in 2013, the MidAmerican Energy Company placed the world's largest onshore wind turbine order to the manufacturer, Siemens Energy. In the same year, the Colorado based energy company Xcel made a major investment in solar and wind energy. As Quin Shea, Vice President of Edison Electric Institute, argues 'You can't simply put the genie back in the bottle when it comes to major strategic investments that the captains of industry are making' (Doniger 2016).

Obama on the international stage

With the enormous economic, political and cultural reach that the US has across the globe, any action or inaction by the US on climate change has major global as well as domestic implications. The potential and importance of the US as a global leader in tackling climate change will be discussed in detail in Chapter 6. But it is important at this stage to provide a brief overview of Obama's role on the international with respect to action on climate change. Sussman and Daynes are rather critical of Obama in this regard. Writing in 2013, they argue that he has had 'several opportunities to demonstrate strong leadership' but 'He has not taken these opportunities' (Sussman and Daynes 2013: 94). Whilst there is some validity in this viewpoint, it does not give us the whole picture. I have already made reference to the speech to the UN early on in his administration when Obama of the necessity for the US to take a leading role in the fight against climate change and the need to support developing nations to mitigate climate change.

Obama's first substantive foray on to the global stage with respect to the issue of climate change was the 2009 UNFCCC, the so-called COP15 held in Copenhagen, Denmark, in December 2009. The UN meeting in Copenhagen was part of a continuing process to get a global agreement on greenhouse gas emission, to follow on from the Kyoto process. However, as Christoff notes 'The Copenhagen COP was mired in controversy almost from its outset' (Christoff 2010). COP15 was marked by bitter division, confusion and setbacks. It was the culmination of intensive negotiations and had a profile unlike any other UN conference on climate change, resulting what was called the Copenhagen Accord. The basis terms of the accord were brokered by President Obama, along with the leaders of Brazil, China, India and South Africa on the final day of the conference. The accord eventually won formal recognition, despite strong opposition from a number of developing countries, including Bolivia, Nicaragua, Sudan and Venezuela, who felt they had been excluded from the negotiations.

Addressing a plenary session of the conference, President Obama argued that the time for talk was over. He went on to argue that 'We can embrace this accord, take a substantive step forward, continue to refine it and to build upon its foundation' (White House 2015). Obama's presence at Copenhagen and his public statements engendered a belief among a number of world leaders that the US now seemed willing to play a positive role on the global stage in tackling climate change, a marked departure from both the tone and the actions of the George W. Bush administration. *New York Times* journalist, Elizabeth Rosenthal, records how in the months leading up to Copenhagen a number of world leaders were becoming more optimistic that the US was now taking the need for global action on climate change more seriously. 'This set off a flurry of diplomacy around the globe' (Rosenthal 2009). Other commentators took a more jaundiced note. Among those was Radoslav Dimitrov, the European Union delegate at Copenhagen. He argued that the conference was a 'failure whose magnitude exceeded our worst fears and that the resultant Copenhagen Accord was a desperate attempt to mask that failure' (Dimitrov 2010).

So what are we to conclude? For President Obama himself the signing of the Copenhagen Accord was 'Ground Breaking' (Sussman and Daynes 2013: 95). Developing the point, Obama argued that Copenhagen was the first time that all of 'the world's major economies have come together to accept their responsibilities to confront the threat of climate change' (Obama 2009). In that sense COP15 could be regarded as partially successful. However, it lacked mandatory commitments. Indeed attempts to give the accord a binding and legal authority failed. It has three broad elements. First, an aspirational goal to limit global temperatures increase to below 2°C (3.6°F) from pre industrial levels. Second, an agreement in broad terms for the reporting of actions by individual countries to reduce greenhouse gases. Third, an outline agreement on a global fund to support developing countries in mitigating climate change. Like so many international agreements, COP15 led to a sub optimal solution, with brokered compromises and package deals. Difficult decisions were left to another day. Subsequent UNFCCC meetings at Cancun (2010), Durban (2011), Doha (2012), Warsaw (2013) and Lima (2014) failed to achieve the substantive breakthrough that those advocating action on climate change had hoped for.

COP21

2015 saw another concerted attempt to broker a global deal on cutting greenhouse gas emission when the nations of the world gathered in December at the Paris UNFCCC, the so-called COP21. The overarching aim of COP21 was for the first to come to a substantive global agreement to tackle climate change involving all the nations of the world. The Paris agreement which came out of COP21 contained a treaty on climate action which included carbon emissions reductions for 187 countries starting in 2020. The treaty came into force when 55 countries covering 55 per cent of global emissions have signed it (COP21 2015).

There were three broad aims: First, to keep global temperature increases well below 2°C (3.6°F), from pre industrial levels and to pursue efforts to keep it below 1.5°C (2.7°F). Second, to review every five years the climate change commitments of the nations of the world at the Paris summit with an initial review in 2018. Third, there was an outline agreement on climate finance for the poorest countries of the world, to support them in the development of renewable energy and clean technologies. This amounts to some $100 billion dollars each year by 2020. This sounds a substantive deal yet to put it in context it represents less than 8 per cent of total declared military spending across the globe.

Individual country commitments were contained in what were tortuously described as intended nationally determined contributions (INDCs). The INDC for the US contained a commitment to reduce its greenhouse gases within a range of 26 to 28 per cent below 2005 levels by 2025. For the American EPA 'The target is fair and ambitious'. It talks of the substantial policy action taken by the US to reduce its carbon emissions. It goes on to argue that 'Additional action to achieve the 2025 target (by the USA) represents a substantive acceleration of greenhouse emissions reductions'. For the EPA 'Substantial emissions reductions are needed to keep the global temperature rise below 2°C and the 2025 target is consistent with a path to deep decarbonisation with the ultimate target of an 80 per cent or more reduction by 2050' (EPA 2016b). Each nation which is party to the agreement has a binding obligation to submit an INDC and agree to a review every five years. However, it is far from clear the binding nature of each individual INDC and the review process itself. Indeed, an assessment published during the COP21 negotiations argued that the INDCs submitted by individual nations would only limit the increase in global temperature to 2–7°C (BBC 2015).

There has also been criticism of the financial support to be given to developing countries to mitigate against climate change. It falls to the green climate fund (GCF) to manage a significant portion of the $100 billion annual fund agreed at Paris. The International Institute for Environment and Development (IIED) has been critical of the GCF arguing that it has a poor record in releasing funds. The IIED has also expressed concern of the danger of prioritising large scale, big ticket, infrastructure projects over more small scale and decentralised projects promoting innovative solutions benefiting people living in poverty – such as off grid energy services for rural communities (IIED 2017). The IIED argues that the GCF needs to support 'entities that can channel funding to reach local communities and build capacity among institutions so that they are agile enough to integrate climate and development initiatives'.

President Obama, however, welcomed the agreement as 'ambitious' and 'historic'. He went on to argue that 'Together we have shown what is possible when the world stands as one'. Whilst recognising the imperfections in the agreement he expressed the view that it was 'the best chance to save the one planet we have' (BBC 2015). The chair of the group representing some of the world's poorest countries argued that 'It is the best outcome we could have hoped for, not just for the least developed countries, but for all the countries of the world' (BBC 2015). However, Nick Dearden, director of the campaign group Global Justice Now argued that the deal 'undermines the rights of the world's most vulnerable communities and has almost nothing binding to ensure a safe and liveable climate for future generations' (BBC 2015). Despite its limitations however, COP21 represented an important step forward on global efforts to tackle climate change.

Conclusion

The challenges facing Barack Obama as the 44th President of the US were huge, complex and varied when he took office in January 2009. The global financial crisis that landed in the autumn/fall of 2008 was sweeping over the US economy, with a major impact on jobs and the housing market. Urgent action in the shape of a stimulus bill was the first priority to ward off a potential economic slump. This had a tendency to crowd out action on climate change, particularly so in the first term of the Obama Presidency The other major challenge facing Obama was the ever growing political partisanship within the US polity, with many Congressional Republicans, often backed by big oil, determined to block the climate change agenda of the President. There is also the broader question of political culture. American society has been built to a large extent on the spirit of individual liberty, where the role of extensive government is viewed with deep suspicion, even downright hostility. Yet the nature of climate change and the challenges it presents require active government solutions. This was always going to be a conflict as Obama sought to break with the Bush years and the roll back on action on climate change and the environment.

Yet in assessing his time in office, Klyza and Sousa have argued that Obama's executive actions on climate change and the environment did make a difference (Klyza and Sousa 2013: 303). Certainly, Obama's engagement with action on climate change and the broader environmental agenda marked a significant departure from the obstructionist policies of the George W. Bush era. Whilst it would be an exaggeration to argue that the US entered a new 'golden age' of environmental policy under the Obama Presidency, there is no doubt that there was substantive shift in the policy debate. Action on climate change and environmental protection became part of the policy mainstream, despite the constraints of the US political system and the strong, sometimes vitriolic, objections of powerful vested interests. It is this interplay of policy making, interests and power that is the subject of Chapter 3.

Notes

1 Renewable energy can be defined as energy that is produced from resources that do not deplete when the energy is harnessed. It includes solar, wind and wave power.
2 The process of hydraulic fracking is highly complex and technical. In broad terms it involves igniting underground explosives to fracture what is known as oil shale. A vertical pipe, often a number of miles deep, is combined with a horizontally drilled pipe which pumps into the shale millions of salty heated water. This is combined with a variety of chemicals to produce a brine under pressure high enough to penetrate the fractures and to release the petroleum and natural gas which is embedded in the shale. The mixture is then captured and pumped to the surface.

References

Bailey, C. (2010) 'Environmental Politics and Policy' in Peele, G., Bailey, C., Cain, B. and Guy Peters, B. (eds), *Developments in American Politics*, Basingstoke: Palgrave Macmillan.

Bomberg, E. and Super, B. (2009) 'The 2008 US Presidential Election: Obama and the Environment' *Environmental Politics*, vol. 18, issue 3, pp. 424–430.

Christoff, P. (2010) 'Cold Climate in Copenhagen: China and the United States at COP15' *Environmental Politics*, vol. 19, issue 4, pp. 637–656.

Cohen, S. and Miller, A. (2012) 'Climate Change 2011: A Status Report on US Policy' *Bulletin of the Atomic Sciences*, vol. 68, issue 1, pp. 39–49.

COP21 (2015) *Conference of the Parties Twenty First Session: Adoption of the Paris Agreement*, 30 November–11 December, Paris.

Dimitrov, R. (2010) 'Inside the UN Climate Change Negotiations: The Copenhagen Conference' *Review of Policy Research*, vol. 27, issue 6, pp. 796–821.

Dunlap, R. McCright, A. and Yarosh, J. (2016) 'The Political Divide on Climate Change: Partisan Polarisation Widens in the US' *Environmental Science and Policies for Sustainable Development*, vol. 58, issue 5, pp. 4–23.

Grunwald, M. (2013) 'Beyond the Keystone pipeline' *Time*, vol. 182, issue 4, pp. 22–24.

Klyza, C. and Sousa, D. (2008) *American Environmental Policy, 1990–2006: Beyond Gridlock*, Cambridge, MA: MIT Press.

Klyza, C. and Sousa, D. (2013) *American Environmental Policy Beyond Gridlock*, Cambridge, MA: MIT Press.

Light, P. (1999) *The Presidential Agenda*, Baltimore, MD: John Hopkins University Press.

MacAskill, E. (2012) 'Bloomberg Backs Obama in Blow to Romney Hopes' *Observer*, 2 December.

Rosenthal, E. (2009) 'Obama's Backing Raises Hope for Climate Pact' *New York Times*, 1 March.

Sussman, G. and Daynes, B. (2013) *US Politics and Climate Change: Science Confronts Policy*, Boulder, CO: Lynne Rienner.

White House (2009) *New Energy Policy for America*, Washington, DC: The White House.

White House (2011) *Blueprint for a Secure Energy Future*, Washington, DC: The White House.

White House (2013) *The Presidents' Climate Action Plan*, Washington, DC: The White House.

White House (2015) *Fact Sheet: President Obama to Announce Historic Carbon Pollution Standards for Power Plants*, Washington, DC: The White House, Office of the Press Secretary, 3 September.

Internet sources

Auel, E. (2016) 'Suing and Spewing: the Massive Pollution Behind the Fight to Overturn the CPP'. Available at: https://cdn.Americanprogress.org/wp-content/uploads/2016/06/22125138/SuingSpewing-brief.pdf, accessed 02/01/17.

BBC (2015) 'COP21 – Climate Change Summit Reaches a Deal in Paris'. Available at: www.bbc.co.uk/news/science-environment-35084374, accessed 14/10/16.

CNN (2014) '5 Reasons Climate Change is Back on the Agenda'. Available at: http://edition.cnn.com/2014/02/18/politics/climate-change-5-things/index.html, accessed 01/10/16.

Deyette, J. (2015) 'EPA Expands the Role of Renewable Energy in the Final Clean Power Plan'. Available at: http://blog.ucsusa.org/jeff-deyette/role-of-renewable-energy-final-clean-power-plan-838, accessed 02/03/16.

Doniger, D. (2016) 'What's next for the Clean Power Plan?'. Available at: www.nrdc.org, accessed 02/02/17.

Durbin, M. (2015) 'Obama's Clean Power Plan will hit Shale Gas Share of Electricity' *Guardian*, 3 August 2015. Available at: www.theguardian.com/environment/2015/aug/03/obamas-clean-power-plan-will-hit-shale-gas-industrys-share-of-energy-generation, accessed 02/08/17.

EPA (2016a) 'Fact Sheet: Overview of the Clean Power Plan'. Available at: www.epa.gov/cleanpowerplan/fact-sheet-overview-clean-power-plan, accessed 16/09/16.

EPA (2016b) 'USA Intended Nationally Determined contributions'. Available at: www4.unfccc.int/submissions/INDC/Published%20Documents/United%20States%20of%20America/1/U.S.%20Cover%20Note%20INDC%20and%20Accompanying%20Information.pdf, accessed 19/11/17.

IIED (2017) 'The Green Climate Fund: Will the Vulnerable be Overlooked in a Rush to Spend?'. Available at: www.IIED.org/green-climate-fund-will-vulnerable-be-overlooked-rush-spend, accessed 02/02/17.

Obama, B. (2009) 'Remarks on Healthcare and Climate Change 19 December'. Available at: www.presidency.ucsb.edu/ws/index.php?=87007, accessed 22/05/17.

Obama, B. (2013) 'Barack Obama Pledges to Bypass Congress to Tackle Climate Change' *Guardian*, 25 June 2013. Available at: www.theguardian.com/world/2013/jun/25/barack-obama-climate-change-strategy, accessed 02/08/17.

Robinson, R. (2016) '24 States are Suing the Federal Government to Stop the CPP – Even Though the Majority of Their Voters Want it'. Available at: www.alternet.org/environment/24-states-are-suing-federal-govt-stop-clean-power-plan-even-though-majority-their-voters, accessed 09/12/16.

3 Interests, influence and power

Introduction

This chapter will argue that the policy making process around climate change in the US is a very crowded one with vested interests on all sides seeking to push their own agendas (for example the oil lobby and environmental pressure groups). Pressure group politics and lobbying are central elements of the policy process in the US. The climate change policy agenda is no exception. There will be analysis of the power and influence of a number of pressure groups across the political spectrum and the ways in which they seek to influence public policy with reference to particular policy issues such as the Keystone XL pipeline. The role of the courts in the policy arena will also be assessed. There will also be an analysis of public opinion and how it shapes the policy agenda. The role of the media in framing the debate around climate change will also come under focus. The analysis in this chapter will be set within the context of the nature of power with a focus on theoretical models of power which seek to explain the impact and locus of power.

Interest group politics, power and influence

Rosenbaum has argued that 'It is an implicit principle in US politics … that organised interests affected by public policy should have an important role in shaping these policies' (Rosenbaum 2014: 44). This may well be so but it is fair to say that some organised interests play a bigger and more influential role than others. Indeed as Gilens argues 'the degree of political inequality in a society, and the conditions that exacerbate or ameliorate it, tell us much about the quality of the society's democracy' (Gilens 2012: 1). Sussman and Daynes talk of 'the extent to which there is congruence between public opinion and policy outcomes remains an important aspect of modern democratic politics' (Sussman and Daynes 2013: 131).

The existence of power and how to measure it is a complex process. Dahl in his classic 1961 study made judgements about who exerted power by analysing actual decisions and their outcomes, in the light of the actors involved and their known preferences (Dahl 1961). Power can also be framed in terms of agenda setting. For Bachrach and Baratz, power is the ability to prevent decisions being made in the first place. This involves the ability of certain groups or individuals to set or control the political agenda, thereby preventing issues from being aired in the first place. Bachrach and Baratz described this as the second face of power (Bachrach and Baratz 1962). Lukes develops this with his idea of the link between hegemony and power. He writes of the third face of power, which is the power of certain groups to render certain ideas or views as illegitimate, thus preventing them even making it on to the political agenda (Lukes 2004).

Power in any political system is far from easy to identify, even in a system as relatively open as the US. Many public policy decisions are often long run and involve a multiplicity of actors and interests. It is therefore difficult to get inside the black box of decision making. How much power and influence any of these groups can exert is dependent on a variety of factors. These can include the financial resources of the group, the size of its membership, the access it has to key decision makers, the way it is portrayed by the media and the support or otherwise of the public.

There are a number of theoretical models which seek to provide an explanatory basis for how power and influence is distributed in a society and political system such as the US. Elite theorists see power as concentrated in the hands of a small number of actors. The Power Elite is a 1956 book by the sociologist, C. Wright Mills. In his seminal work, Wright Mills writes of the concentration of power in US society, within a network of the leaders of the military, corporate and political elements of society, sharing mutual and interwoven interests (Wright Mills 1956). In similar vein, writing in the 1970s, David Riesman argued that power in the US had fallen into the hands of a variety of what he termed 'veto groups', with each group pressing to influence the policy agenda.

'There are, of course still some veto groups that have more power than others' argued Riesman 'But the determination of who these are has to be made all over again for our time' (Riesman 1970: 26–30). What was true on 1970, it can be argued, is equally true in 2017.

In contrast to elite theory, pluralism in its various early forms saw power as less concentrated with a range of actors seeking to influence the policy agenda. Not all actors are equal, business and corporate interests are often to the fore, but a range of competing interests result in varying degrees of democratic equilibrium (Dahl 1961; Newton 1976). However, some pluralists such as Lindblom refined their views over time. In his 1977 work, *Politics and Markets*, he noted the privileged position of business within society, with the result that real choices and competition among various actors are limited, thus echoing some of the central observations of elite theory (Lindblom 1977).

In the US there are a range of groups on differing sides of the political debate who seek to influence the policy agenda on climate change and the environment more generally. The pressure group arena nested around the issue of climate change is a crowded one with an ever increasing number of interests coming to the fore. On one side of the argument are what Sussman and Daynes classify as a group of 'contrarians or deniers' who continue 'to aggressively resist the science'

(2013: 132). These include conservative and right wing policy think tanks such as the Cato Institute, the Heartland Institute, the Heritage Foundation and other organisations representing big oil and the coal sector. Opposing them is the environmental lobby. This lobby is far from homogenous, representing as it does a broad range of groups. They include the Sierra Club, Friends of the Earth, Green-peace, the NRDC, the Clean Energy Coalition and the National Wildlife Federa-tion. There are also a number of evangelical Christian groups active in the environmental lobby that are influenced by biblical teachings on the stewardship of the earth. Examples include Christians and Climate and Creation Care.

Power and influence are not always easy to identify, even in a relatively open political system such as the US. There are also multiple points of access in the US political system which may provide avenues for a variety of groups to influ-ence policy. Central to the US constitution was a desire by the founding fathers to prevent an over powerful and centralised state. The doctrine of the separation of powers gives specific authority to each of the three branches of federal gov-ernment, the Presidency, the Congress and the Supreme Court. The individual states also have their own areas of autonomy and influence laid out in the constitution.

In the US, as Rosenbaum notes, 'Arrangements exist throughout govern-mental structures for giving groups access to strategic policy arenas' (Rosen-baum 2014: 45). Indeed lobbying by interests groups in the US is regarded as a key element of the public policy arena and the democratic process itself. As an illustration of this, there are over 1000 advisory committees within the federal level of government, the purpose of which is to give interest groups some access to the decision making process. But how does such access pan out in terms of the power and influence of interest groups? This takes us back to our earlier theoretical discussion about the distribution of power in political systems.

Rosenbaum argues that with regard to environmental policy, 'No interest has exploited the right to take part in the governmental process more persuasively or successfully than has business' (Rosenbaum 2014: 45). Such a view is reflective of the elite power theory. The business sector (in particular the large corporate sector) has a number of key advantages when it comes to influencing govern-ment over other groups. In general, it is well resourced with an organisational structure that can respond quickly and effectively to take action when its inter-ests are threatened. Business has as a 'special relationship with government' (Lindblom and Woodhouse 1993: 90). Echoing this view point, Klein talks of the 'corporate owned media' and how it tries to block out alternate voices which might threaten dominant interests (Klein 2014: 369). In this regard, Kingsworth argues that 'The notion that environmentalists are a privileged elite telling the hard pressed that they can't have decent lives has become a staple of corporate propaganda for decades' (Kingsworth 2017).

However, does this apparent validation of elite power theory reflect the total-ity of interest group activity with regard to environmental policy and action on climate change, specifically the role of the environmental lobby? In this regard, Rosenbaum records how since the 1970s a series of structural arrangements have been put in place by both federal agencies and the US Congress. He argues that these have opened up avenues of influences for environmental lobby groups, which allow them to shape and influence policy on climate change and the environment more generally (Rosenbaum 2014: 46). The environmental lobby has made use of the courts to enhance action on climate change and to delay or block controversial infrastructure projects. I will return to the role of the courts subsequently.

Such examples might be more reflective of the pluralist perspective on interest group activity. However, a note of caution is needed here. Public policy making is a complex, detailed and long term process with multiple points of access and influence. As I noted earlier, even in a country such as the US with its relatively open form of government, it is far from easy to get a handle on who has the most impact on policy decisions. If Lukes is correct about how the powerful vested interests can stop issues even being debated in any substantive form, the so-called third face of power, then the situation becomes even more opaque (Lukes 2004). And yet available evidence would suggest a real resource imbalance between the environmental lobby and corporate interests such as big oil. For example, in the first year of the Obama administration, oil and gas interests spent more than $82m lobbying Congress on climate change legislation compared to just under $1m by the environmental lobby.

If we go back to the George W. Bush Presidency, in June 2005 released papers from the US State Department showed the administration thanking Exxon oil executives for the company's active involvement in helping to determine climate change policy (Vidal 2005). This example is part of a broader pattern of how what can be loosely termed the climate science denial movement who use their influence to attack the science on climate change and to impact on the policy agenda. In this regard, former US Vice President, Al Gore has written about the 'brazen efforts' of the oil giant Exxon Mobil 'to try to manage public perceptions of the reality and seriousness of climate change' (Gore 2008: 201).

Friedman has developed a typology of climate deniers (2008). First, there are those resourced by fossil industry think tanks such as the Cato Institute and the Heritage Foundation. Second, there are the very small minority of scientists who draw differing conclusions from the data, arguing that increases in global temperature are due to natural climatic variability. Third, there are conservative Republicans and supporters of the Tea Party who refuse to accept the reality of climate change because what it would necessitate, more government regulation intervention. This goes against all their ideological beliefs.

For the journalist and campaigner Naomi Klein, the 'ties between the deniers' and big oil and corporate interests are well known. For example, the Heartland Institute, a policy think tank, has received in excess of one million dollars from Exxon Mobile and other corporate interests (Klein 2014: 44). A study in 2013 by Dunlap and Jacques found that 72 per cent of climate change denial books, most of which were published since the 1990s, were linked to Conservative right wing think tanks (Dunlop and Jacques 2013). The sociologist Robert Brulle has labelled such think tanks 'the climate change counter movement'. He calculates that they draw in some $900m a year from Conservative backers that cannot be fully traced (Brulle 2014). For Klein, the overall aim of the climate change deniers is to protect 'powerful political and economic interests' (Klein 2014: 44). Klein's analysis is reflective of the elite theory of power. Klein's position can be characterised as on the liberal left of the American spectrum. She seeks to put forward a new concept of power, what she terms the politics of human power. She poses the question as to whether there can be a shift from corporations towards communities. For Klein it is a 'shift that will require rethinking the very nature of humanity's power' (Klein 2014: 25). In essence what Klein is talking about here is the constant merry-go-round of consumerism and the rapacious use of the finite resources of planet earth and the need to move towards a more sustainable way of living. This would require not only a shift in power relations but a change in the mindset of both political decision makers and citizens alike. There is no doubting that this is a major challenge but the exigencies of climate change and the threat it poses to people and planet means that it is a challenge that cannot be ducked.

The role of the courts

As I discussed in Chapter 1, at the federal level the US is governed on the constitutional principle of the separation of powers with the key institutions in Washington, DC being the President, the Congress and the Supreme Court. All three branches of government play a key role in policy making. The focus here is on the role of the courts in shaping environmental and climate change policy. The US Supreme Court is at the apex of a complex system of courts in the US. I have already made reference in the previous chapter to the controversial decision to delay the implementation of the Obama administration's CPP. As Klyza and Sousa note these courts have always played a key role in environmental policy but over the last decade or so 'continuous legislative deadlock has increased the significance, complexity and the contentiousness of the judicial pathway' (Klyza and Sousa 2013: 141). Interested parties across the political spectrum with differing views about action on climate change and environmental policy more generally, use the courts to get results which are difficult to achieve in the policy arena due to the legislative gridlock in Washington, DC. As a consequence, 'The courts are likely to remain important environmental policy makers' (Klyza and Sousa 2013: 176).

Since the early beginners of the environmental movement in the 1960s, environmental lobby groups have used action in the courts as a key element in their strategy to protect and improve the environment. Whether it is the 1973 Endangered Species Act, air and water pollution, mining on federal lands, and the disposal of hazardous waste, the courts have played a pivotal role in shaping and regulating policy. 'Therefore it should come as no surprise,' as Engel notes, 'that many advocates of action on climate change are turning to the courts' (Engel 2010: 229). The courts have become an ever important locus of action on climate change over the last decade, with vested interests on various sides of the argument fighting their particular corner. Engel has identified three broad strategies pursued by litigants. First, actions to compel federal agencies to regulate greenhouse gases under existing legislation, the 2007 case of Massachusetts versus the EPA being a case in point. I will return to this shortly. The second strategy involves pushing federal agencies to take into account action on climate change when making particular decisions, for example the building of new coal fired power stations. The third line of action involves attempts to hold to account utilities, energy companies and the like for their greenhouse gas emissions under the common law.

As Engel notes, the first of these three strategies is the one most commonly employed (Engel 2010: 232) and tends to involve a coalition of states and environmental lobby groups. California, Connecticut, Massachusetts, New York, New Jersey and Oregon are among the states most active in climate policy and pursuing action through the courts. These states often appear as plaintiffs with high profile environmental groups such as the Centre for Biological Diversity, the NRDC and the Sierra Club. Most of the examples involve so-called second tier cases, that is in courts below the Supreme Court. These include the Ninth Circuit of Appeals.

But a notable example involving the Supreme Court is the 2007 legal case of Massachusetts versus the EPA. For Sussman and Daynes this case was of particular importance because it opened the door for court action on climate change in the twenty-first century (Sussman and Daynes 2013: 111). The case had its genesis in 2003 when the EPA, at the time of the G. W. Bush administration, rejected a petition demanding that the Federal government and agencies restrict emissions of greenhouse gases, principally CO_2 using the authority of the Clean Air Act. The EPA rejected the petition arguing that greenhouse gas emissions did not constitute an air pollutant and were therefore outside the realm of the 1970 Clean Air Act. In 2006, 12 US states, 3 local government jurisdictions, 13 environmental lobby groups and 6 former directors of the EPA took their case to the US federal Supreme Court and appealed the decision of the US Court of Appeals which had upheld the EPA stance. Environmental groups during the Bush Presidency regarded the Supreme Court with suspicion when it came to environmental laws. However, in a judgement reached as the G. W. Bush Presidency was drawing to an end, it found in favour of the plaintiffs by a majority of five to four. In a ruling that was highly critical of the approach of the EPA to air quality during the G. W. Bush administration, it issued a judgement that the Clean Air Act gave the EPA authority to regulate greenhouse gases. It was an outcome that caused surprise in some quarters. Indeed as Rosenbaum argues, the 'Supreme Court has seldom been wholly predictable about environmental issues' (Rosenbaum 2014: 114).

However, the majority verdict reflected an ideological division within the Supreme Court between liberals who favour action on climate change and conservative judges who take a different view of the Court's role. It is a division that has been a feature of a number of decisions on a range of issues over decades and has had implication for climate action under the Obama administration. A second case involving greenhouse gas emissions taken to the Supreme Court was that of American Electric Power versus Connecticut (2011). Principle litigants in the case were the states of California, Iowa, New York, New Jersey, Rhode Island, Vermont and Wisconsin. The states sought action against six energy companies over polluting emissions in their respective states. The energy companies in the spotlight included American Electric Power, Cinergy Corporation and Xcel Energy. The case was that the emissions produced by the six energy companies amounted to a 'public nuisance' and as a consequence they should cap and reduce over time their CO_2 emissions (Sussman and Daynes 2013: 114). After two months of deliberation the Court by a majority of eight to zero with one abstention rejected the argument that carbon dioxide emissions could be capped using the standard of 'public nuisance'. Instead it reaffirmed the 2007 Massachusetts versus EPA decision that federal action on such emissions was to be based on the actions of the EPA under the provisions of the Clean Air Act. The judgement did cause concern amongst a variety of environmental groups. However, an editorial in the *New York Times* struck a more optimistic tone arguing that the decision was a 'clarifying and positive decision vindicating the Clean Air Act' as a legal basis for government action, specifically the EPA, on the control of greenhouse gases (*New York Times*, The Carbon Ruling, 21 June 2011). It was an important decision in the context of an increasingly divided polity which by 2011 had resulted in a virtual legislative deadlock as regards action on climate change in the Congress. In 2016, as I have already noted in Chapter 2, the Supreme Court was again called into action when a legal challenge was made to a flagship policy of the Obama Presidency on climate change, namely the CPP.

These three upper tier cases involving the Supreme Court are clearly of significance. In addition, however, there are a number of second tier cases with regard to environmental and climate change policy, that is litigation in courts sitting below the level of the Supreme Court. One such case is that of Comer versus Murphy Oil (2009). It was brought by a group of people who suffered property damage as a result of Hurricane Katrina. Murphy Oil and other energy companies were charged with various violations that the litigants claimed increased the severity of the hurricane. These included public nuisance, negligence and trespass. The case was dismissed and the attempt by the litigants to appeal to the Supreme Court was denied. Another case where the litigants tried to roll back regulations to control greenhouse gases was that of the Chamber of Commerce of the United States of America and National Automobile Dealers Association (NADA) versus EPA (2011). The litigants challenged the authority of the EPA to allow individual states to have stricter vehicle emissions standards than those at the federal level. The appellate court that sat in judgement dismissed the case arguing that the emissions standards did not have a direct impact on the Chamber of Commerce and the NADA.

One important issue which has had the attention of the courts over several years in relation to climate change and environmental policy is that of environmental justice. Whilst there is no clear consensus as to its exact meaning, the EPA has defined environmental justice as 'the fair treatment and meaningful involvement of all people regardless of race, color, national origin or income with respect to the development, implementation or enforcement of environmental law, regulation and policies' (EPA 2004) The Environmental Justice Foundation (EJF) defines environmental justice in broad terms as protecting the environment and human rights (EJF 2017).

The principle legal route for environmental justice cases are local and state courts. However, as Rosenbaum argues, despite some judgements in favour of plaintiffs claiming significant harm as a consequence of environmental injustice, such local and state courts 'have not proven to be a major venue for the environmental justice movement' (Rosenbaum 2014: 153). The key to this is the lack of an effective statutory basis across the majority of localities and states. Indeed only six states have put in place legislation creating a legally enforceable right for seeking redress for environmental injustice. The federal courts have proved a more successful avenue for cases involving environmental injustice. The key basis for legal action is Title VI of the 1964 Civil Rights Act which allows cases to be brought to court involving discrimination against ethnic, racial and religious minorities and women. Advocates of environmental justice have achieved modest but nonetheless substantive progress enforcing civil rights. Examples include action by the US State Department on halting human trafficking.

One important legal case regarding environmental justice is that of Native Village of Kivalina versus Exxon Mobil (2009). The case concerns a remote and isolated Alaskan village with a population of 400 people. Rising temperatures have led to a reduction in ice in the area, threatening their livelihoods and leaving the village vulnerable to extreme weather. The Army Corps of Engineers has estimated that the village will need to relocate at a cost of up to $125 million. The village brought Exxon Mobil to court as the principal offenders against the welfare and health of its people. The case was originally taken to the US District Court for Oakland but was dismissed. The village then appealed to the Ninth Circuit Court of Appeals (Washington, DC). However, the court ruled that the federal common law was not available to resolve the issues raised by the case of Kivalina. Only the legislative and executive branches could provide a remedy. However, in a further case in 2014, the Ninth Circuit Court of Appeals ruled in favour of native Alaskan tribes, supported by a number of environmental groups and ordered Shell Oil to stop its exploratory drill in the Arctic waters off Alaska for fear of an oil spill.

To sum up, while Sussman and Daynes are right to argue that 'the judiciary has not been the primary actor in the United States in responding to global climate change' (Sussman and Daynes 2013: 125), it is has nonetheless being an important factor in shaping US environmental and climate change policy.

The media

Smith has argued in the context of the US that 'The media help shape what people think, feel, and do about environmental issues' (Smith 2014: 121). From the *New York Times* articles on the hazards of x-rays in the early 1900s, television documentaries on population and conservation in the 1970s and 1980s up to digital and social media representation of climate change today 'the quality of media story telling has informed the nature and extent of public and political responses' (Smith 2014: 121). Whilst it is certainly correct to argue that the media can play an important role in shaping perceptions and shaping policy on issues such as climate change, it is also important to focus on the quality and accuracy of its reportage. This is particular so with regard to climate change. In the 1980s climate change was not the partisan issue that it has become over the last decade. As Chong notes, 'Scientists were the predominant group speaking on the issue in the media and their scientific reports became the basis of news stories that set the agenda' (Chong 2015: 123). This can no longer to be said to be the case as media coverage has taken on an increasingly ideological hue. There is also a more general problem with media coverage of climate change. Climate change is a wicked and complex problem. This has implications for how it is covered by the US media. Indeed as Boykoff has argued 'issues associated with climate change become more challenging and complex to cover as understanding of its many dimensions develops' (Boykoff 2011: 168).

Print journalism has, for example, over the last decade faced severe financial pressure with the loss of revenue from sales, advertising and the like. As Smith acknowledges, as a consequence newspapers have seen a reduction in their ability to recruit specialist journalists. The consequence of this, Smith argues, is that 'investigative journalism has become an increasingly rare practice' with journalists having 'little time to research the background to complex stories' (Smith 2014) And as we have seen, climate change is one of those immensely complex stories.

A new word has entered the lexicon of print journalism in recent years – that word is 'churnalism'. It describes a situation where newspapers shorn of journalistic talent, reproduce press releases with minimum alteration. This can play into the hands of well resourced powerful corporate interests such as big oil and is reflective of the elite model of power that we discussed earlier on in this chapter. Oreskes and Conway's survey of corporate power is a case in point. Based on their research, they argued that there was a close interconnection between corporate interests representing tobacco and big oil and media outlets in relation to the health impacts of tobacco and the causes and impact of climate change which had consequences for the way these issues were reported (Oreskes and Conway 2010). But to what extent is this portrayal of such concentration of influence a fair and accurate reflection of media/interest group relations? Smith, for example, argues that well established environmental NGOs, such as the Sierra Club, the NRDC and Friends of the Earth, can also have influence over hard pressed print journalists, anxious to fill space However, Tegelberg *et al.* argue that in the US mainstream press coverage on the subject of climate change has been on the decline. They note that in 2007 the release of the IPCC's Fourth Assessment Report and the release of Al Gore's documentary movie *An Inconvenient Truth* sparked significant mainstream media interest in the causes of climate change and the policy implications (Tegelberg *et al.* 2014: 67). But as the authors note in the subsequent decade, 'coverage has steadily diminished' in part a product of 'aggressive efforts to discredit climate science' (Tegelberg *et al.* 2014: 68). Sussman and Daynes make a similar point arguing that the 'vocal minority of sceptics and deniers of climate change have been given more coverage than merited by their numbers' (Sussman and Daynes 2013: 155). Naomi Klein makes reference to all the news stories that were never published or aired with regard to climate change. This, she argues, reflects the power of big oil and other corporate lobbyists (Klein 2014: 34). Such a view is reflective of the elite view of power and the concept of the third face of power that I discussed earlier in this chapter. Whilst such a viewpoint is open to challenge, evidence does point to a reduction in climate change coverage in some media outlets. In 2007, for example, the three major US networks, CBS, NBC and ABC, ran 147 stories but in 2011 there were just 14 (Klein 2014: 34). Research by the organisation Media Matters found that the same three networks aired only 50 minutes coverage in the whole of 2016 (Kahlhoefer *et al.* 2017). Research by the Yale programme on climate change communication in 2016 found that only 44 per cent of Americans say that they hear global warming discussed once a month or more (Yale 2016a).

A key challenge for the media in its reportage of climate change is getting the balance right between competing claims and conflicting opinions. Do the media give too much weight to both sides of the argument when the balance of the evidence is so clearly on one side, that climate change is happening and it is a consequence of human action? A 2004 study by two political scientists gives us an interesting into this aspect. It involved an analysis of the content of stories about climate change and global warming in the *New York Times*, the *Washington Post*, the *Los Angeles Times* and the *Wall Street Journal*. More than half of the news stories gave almost equal attention to the science on climate change and to what can only be described as the fringe position that global warming is not caused by human action. Only 35 per cent of the stories gave more weight to the dominant scientific view (Boykoff and Boykoff 2004). The mainstream media is often criticised by scientists for a creating a 'climate of doubt' by giving too much weight to those who challenge the science, leading to public confusion. In this regard Chong talks of so-called media objectivity 'in the sense of presenting opposing arguments without regard to their scientific status … giving audiences a distorted picture of reality' (Chong 2015: 124).

Partisanship and the media

Painter has argued that 'The US media coverage of climate change has been both shaped and being shaped by' the high level of political polarisation in the US (Painter 2013: 127). The abolition of the Fairness Doctrine in 1987 which required broadcast companies to be 'honest, fair and balanced' in their coverage of controversial issues has opened up a new media terrain with clearly politically affiliated TV broadcasters such as Fox News and MSNBC. Fox News follows a conservative agenda and is well known for pushing the climate sceptic agenda.

Its high profile conservative commentators have included Pat Buchan and Sarah Palin. MSNBC's political sympathies are on the liberal left. It has a very different take on climate change from that of Fox News. Notable presenters include Rachel Maddow and the Reverend Al Sharpton.

In relation to the print news media, research evidence shows that there is significantly more space given to climate change sceptics and deniers. Painter and Tickell-Painter in their study of print media coverage of climate change looked at three newspapers: the *New York Times*, *USA Today* and the *Wall Street Journal* (Painter 2013). These three newspapers have the highest circulations in the US and play an important role in agenda setting. The *New York Times* is regarded as a paper with a liberal stance with a reputation for covering environmental issues such as climate change. Indeed it set up a dedicated environment news desk in 2009. However, it was subsequently closed in 2013. Nonetheless, it continues to cover issues around climate change. The *Wall Street Journal* follows a broadly conservative agenda. It has given space to the views of climate change sceptics such as Rush Limbough and Bjorn Lomborg. Of the three newspapers in the study, the *New York Times* gave the most consistent space to climate change, while *USA Today* gave the least.

The role of social media

In January 2009, the Pew Research Center's Project for Excellence in Journalism began to monitor the weekly news coverage in the US, drawing a distinction between traditional coverage (TV, newspapers and radio) and new social media. Its analysis shows that issues around global warming and climate change have received a much larger share of coverage in the new social media such as Facebook and Twitter. Boykoff asks whether the increased coverage of climate change in social media translates into 'improved communication or just more noise' (Boykoff 2011: 169). There are a number of differing views in respect of this question. Jones, for example, has argued that 'good journalism on the web is a wondrous thing' which has produced a 'ferment of creative journalistic thinking (Jones 2009: 179). Sunstein paints a more negative picture of the value of social media, warning of the likelihood of the 'echo chamber' effect whereby those online simply consume news and opinion that are in accord with their particular world view (Sunstein 2007). Chilsom makes a similar point to Sunstein. He notes that many social media posts on global warming and climate change are linked to stories in the mainstream media. Often these are 'stories that support the individual's personal belief about whether global warming is a concern or not' (Chilsom 2014).

For Tegelburg *et al.*, 'The pervasiveness and widespread influence of mobile communication devices, combined with limited news budgets help explain the decline in offline climate news and a corresponding rise in on line coverage' (Tegelburg *et al.* 2014: 68). Research by the Pew Research Centre shows a media landscape with more and more Americans getting their news online, shifting away from traditional sources of news and information, such as the printed press and television. When Pew began tracking social media adoption in 2005, it found that just 5 per cent of American adults used at least one of these platforms. By 2011, that share had risen significantly to 50 per cent. By 2016, 69 per cent of the American public was using some form of social media (Pew Research 2017).

There is no doubt that there is long term trend in the use of online news sources, through the use of social media. However, what is less clear is the impact of social media with regard to climate change. How many people are engaging with the issue of climate change on social media outlets such as Facebook and Twitter? Is this debate cutting through to the general public or it just an inward facing debate amongst the already committed? One possible way to measure the impact of social media and the media in general is to survey public attitudes to climate change and its importance (or not) in the collective consciousness. It is to this that I now turn.

Public opinion, the environment and climate change

For Sussman and Daynes, 'An assessment of public opinion polls over the last decade reveals that Americans generally support environmental protection' (Sussman and Daynes 2013: 143). And yet a word of caution is needed here. A majority of Americans may support environmental protection but, when asked in opinion poll surveys, it is an issue which generally ranks quite low in citizens' policy priorities. In other words, environmental protection lacks policy saliency for many American voters. As a consequence 'assessing Americans' attitudes to the issue of environmental protection is somewhat difficult given this dichotomy' (Sussman and Daynes 2013: 143). When it comes to the specific question of climate change, public opinion is complex and multifaceted. Despite the overwhelming consensus in the climate science community in the US and globally about anthropogenic (human induced) climate change, American public opinion is much more divided, both on the actual incidence of climate change, and the causes behind it.

In December 2011, Pew Research published data about public opinion and climate change. It showed that the proportion of Americans who believed that there was 'solid evidence' that the earth was warming dropped from 77 per cent in 2006 to 57 per cent in 2009. Such a drop is perhaps reflective of the increasingly ideological nature of the climate change debate, especially since the election of Barak Obama in 2008. It might also be in part a product of the efforts of the climate change denial lobby that I have referred to in this chapter. Pew's data for 2011 showed a small increase to 63 per cent but still significantly lower than the figure for 2006 (Pew Research 2011).

A key issue with regard to climate change is its policy saliency. In other words, what priority does the American public give to tackling the issue? In November 2015 Pew Research published the findings of comparative global analysis of public attitudes to climate change. It found that only 45 per cent of Americans regard climate change as a very serious issue. This compared to 54 per cent in Europe,

74 per cent in Latin America, and 61 per cent in Africa (Pew Research 2015). The same survey also showed a stark partisan divide on the issue of climate change in the US. 68 per cent of Democrats thought that global climate change was a very serious problem compared to just 20 per cent of Republicans. Linked to this, 82 per cent of Democratic party supporters supported action to limit greenhouse gases compared to 50 per cent amongst Republicans (Pew Research 2015). The latest data reinforces this partisan divide. In 2016, in the run up to the Presidential election, Pew Research surveyed registered supporters of the Democratic Presidential candidate, Hillary Clinton and the Republican candidate, Donald Trump. It found that 56 per cent of Clinton supporters cared a great deal about climate change compared to just 15 per cent of Trump supporters. With regard to the causes of climate change, 70 per cent of Clinton supporters believed it to be the result of human activity, for supporters of Donald Trump, the figure stood at 22 per cent (Pew Research 2016b).

What do voters overall in the US think about the causes of climate change? Recent research by Pew found that 50 per cent of all voters believed that climate change was the result of human activity (Pew Research 2016b). The Yale programme puts the figure at some 53 per cent (Yale 2015).

What are then some possible explanations for this disconnect between the scientific evidence on the incidence and causes of climate change and public attitudes? As we have already noted, climate change has been described as a wicked problem. It is complex and difficult to understand and it lacks immediacy in terms of its impact for many Americans. The public in the US, or at least a significant section of it, view action to tackle climate change as a long term and distant objective, not a policy priority for the short and medium term. Furthermore, since the global financial crisis of 2007/2008, the economy, lack of job prospects and general financial insecurity have much more salient policy issues than the environment and climate change for many Americans, especially those in the so-called 'middle class'.

Some commentators argue that the confusion amongst the American public about climate change is the product of a well funded campaign by corporate interests and policy think tanks, including the Cato Institute, the Heritage Foundation and Exxon Mobil, who actively promote a climate change sceptic agenda (Klein 2014; Gore 2008). More broadly, the issue of climate change in the US has become much more politicised and ideological in the last decade, with one's political party identity seeming to be an important determining the attitude of voters towards climate change. Indeed for Sussman and Daynes, 'ideology has become the prism through which supporters and opponents view climate change' (Sussman and Daynes 2013: 146). And therein lies the problem. Such political polarisation poses a challenge to those attempting to build support for substantive action on climate change.

However, some of the latest data does seem to point to something of a shift in public opinion. For example, research by the Yale Program on Climate Change Communication in the Spring of 2016 found that 70 per cent of Americans thought that global warming was happening, an increase of 7 per cent on the previous year. This is the highest figure recorded since November 2008 when it stood at 71 per cent, in a period just before the financial crisis took hold across the US (Yale 2016b). In the same Yale survey, 38 per cent of those interviewed thought that people in the US were being harmed right now by climate change, up from 32 per cent in 2015. Furthermore, a Gallup poll in March 2016 found that concern about global warming had increased among all party groups since 2015. Overall, 64 per cent of Americans are worried a great deal or a fair amount about global warming, up from 55 per cent in 2015, and the highest figure recorded since 2008. Within the figures there is still evidence of an ideological divide. Amongst Democrats the figure stands at 84 per cent, up 6 per cent on the previous year. Whereas only 40 per cent of Republicans are worried a great deal or a fair amount about global warming, but this up by a significant 9 per cent on the previous year. In the same survey, 65 per cent of Americans said that increases in temperature were primarily attributable to human action rather than natural causes (Saad and Jones 2016).

Whilst one needs to be cautious in drawing firm conclusions from these findings, they do seem to suggest something of a shift in public attitudes to global warming and climate change and to its policy saliency, at least in the short term. Saad and Jones make the point that factors such as the economic downturn in the US economy (a product of the 2008 financial crisis), and the activities of the climate change denial lobby 'may have dampened public concerns' from 2009 to 2015. However, the cumulative effect of unusually warm weather in the years 2011 to 2016, coupled with a number of severe weather events such as hurricane Sandy in 2012 have 'potentially contributed to this shift in public attitudes' (Saad and Jones 2016). We will need to see more data on this in the coming years to see whether this is part of a long-term trend.

Conclusion

This chapter has focused on the interplay between interests, influence and power in the politics of climate change. This involves a range of actors on different sides of the political debate, ranging from environmental lobby groups to those seeking to undermine the science of climate change. There was also a focus on the impact of the media on the public discourse around climate change. The key role of the courts in shaping the policy agenda on climate change and the environment has also been an important area for analysis. How such groups and institutions influence the policy agenda is contingent on the power and influence they can bring to the table. The extent to which the climate change policy arena is a relatively closed one, open to only a small circle of corporate interests, or a more open pluralist space open to a range of actors is, as we have seen, a matter for much debate. In the next chapter, I turn my attention to a key aspect of the climate change policy debate, namely the politics of energy policy.

References

Bachrach, P. and Baratz, M. (1962) 'Two Faces of Power' *The American Political Science Review*, vol. 56, issue 4, pp. 947–952.

Boykoff, M. (2011) *Who Speaks for the Climate? Making Sense of Media Reporting on Climate Change*, Cambridge: Cambridge University Press.

Boykoff, M. And Boykoff, J. (2004) 'Bias as Balance: Global Warming and the US Prestige Press' *Global Environmental Change*, vol. 14, issue 2, pp. 125–136.

Brulle, R. (2014) 'Institutionalising Delay: Foundation Funding and the Creation of the US Climate Change Counter Movement Organisations' *Climate Science*, vol. 122, issue 4, pp. 681–694.

Chong, D. (2015) 'Explaining Public Conflict and Consensus on the Climate' in Wolinsky-Nahmias, Y. (ed.), *Climate Change Policies: US Policies and Civic Action*, London: Sage.

Dahl, R. (1961) *Who Governs? Democracy and Power in an American City*, New Haven, CT: Yale University Press.

Dunlop, R. and Jacques, J. (2013) 'Climate Change Denial Books and Conservative Think Tanks: Exploring the Connection' *American Behavioural Scientist*, vol. 57, issue 6, pp. 699–731.

Engel, K. (2010) 'Courts and Climate Policy: Now and in the Future' in Rabe, B. (ed.), *Greenhouse Governance: Addressing Climate Change in America*, Washington, DC: Brookings Institution Press.

Friedman, T. (2008) *Hot, Flat, and Crowded*, New York: Farrar, Straus and Giroux.

Gilens, M. (2012) *Affluence and Influence*, Princeton, NJ: Princeton University Press.

Gore, A. (2008) *The Assault on Reason*, New York: Penguin Books.

Jones, A. (2009) *Losing the News: The Future of the News that Feeds Democracy*, Oxford: Oxford University Press.

Kingsworth, P. (2017) 'The Lie of the Land' *Guardian*, 18 March.

Klein, N. (2014) *This Changes Everything: Capitalism vs the Climate*, London: Allen Lane.

Klyza, C. and Sousa, D. (2013) *American Environmental Policy Beyond Gridlock*, Cambridge, MA: MIT Press.

Lindblom, C. (1977) *Politics and Markets*, New York: Basic Books.

Lindblom, C. and Woodhouse, E. (1993) *The Policy Making Process*, Englewood Cliffs, NJ: Prentice Hall.

Lukes, S. (2004) *Power: A Radical View*, Basingstoke: Palgrave Macmillan.

Newton, K. (1976) 'The Theory of Pluralist Democracy' in McGrew, A. and Wilson, M. (eds), *Decision Making: Approaches and Analysis*, Manchester: Manchester University Press.

Oreskes, N. and Conway, E. (2010) *Merchants of Doubt: How a Handful of Scientists Obscured the Truth on Issues from Tobacco Smoke to Global Warming*, New York: Bloomsbury Press.

Painter, J. (2013) *Climate Change in the Media: Reporting Risk and Uncertainty*, London: I. B. Taurus.

Riesman, D. (1970) 'The Veto Groups' in Crockett, N. (ed.), *The Power Elite in America*, Lexington, MA: DC Heath and Company.

Rosenbaum, W. (2014) *Environmental Politics and Policy*, Los Angeles: Sage.

Smith, J. (2014) 'Commentary on Part 11: Climate Change Media' in Crow, D. and Boykoff, M. (eds), *How Information Shapes our Common Future*, London: Routledge Earthscan.

Sussman, G. and Daynes, B. (2013) *US Politics and Climate Change: Science Confronts Policy*, Boulder, CO: Lynne Rienner Publishers.

Sunstein, C. (2007) *Republic.com 2.0*, Princeton, NJ: Princeton University Press.

Tegelburg, M., Yagodin, D. and Russel, A. (2014) 'Climate News, Summit Journalism and Digital Networks' in Crow, D. and Boykoff, M. (eds), *How Information Shapes our Common Future*, London: Routledge Earthscan.

Wright Mills, C. (1956) *The Power Elite*, Oxford: Oxford University Press.

Internet sources

Chilsom, M. (2014) 'Define Global Warming, Twitter: Social Media Take on Climate Change'. Available at: www.newsmax.com/FastFeatures/define-global-warming-twitter-social-media/2014/10/31/id/604304/, accessed 03/11/16.

EJF (2017) Available at: www.ejfoundation.org, accessed 04/03/17.

EPA (2004) Available at: www.epa.gov./compliance/environmentaljustice, accessed 04/03/17.

Kahlhoefer, K., Seifter, A. and Robbins, D. (2016) 'How Broadcast Media Covered Climate Change in 2015'. Available at: www.mediamatters.org/research/2016/03/07/study-how-broadcast-networks-covered-climate-ch/208881, accessed 23/03/17.

Pew Research (2011) 'Modest rise in number saying there is solid evidence of global warming'. Available at: www.people-press.org/2011/12/01/modest-rise-in-number-saying-there-is-solid-evidence-of-global-warming/, accessed 02/02/17.

Pew Research (2015) 'Global Concern about Climate Change, Broad Support for Limiting Emissions'. Available at: www.pewglobal.org/2015/11/05/global-concern-about-climate-change-broad-support-for-limiting-emissions/, accessed 02/02/17.

Pew Research (2016a) Available at: www.pewresearch.org/topics/energy-and-environment/, accessed 02/02/17.

Pew Research (2016b) Available at: www.pewresearch.org/fact-tank/2016/10/10/clinton-trump-supporters-worlds-apart-on-views-of-climate-change-and-its-scientists/, accessed 02/02/17.

Pew Research (2017) 'Pew Research Social Media'. Available at: www.pewresearch.org/topics/energy-and-environment/, accessed 04/04/17.

Saad, L. and Jones, J. (2016) 'US Concern About Global Warming at Eight-Year High'. Available at: www.gallup.com/poll/190010/concern-global-warming-eight-year-high.aspx, accessed 21/03/17.

Vidal, D. (2005) 'Revealed: How Oil Giant Influenced Bush'. Available at: www.theguardian.com/news/2005/jun/08/usnews.climatechange, accessed 03/10/16.

Yale (2015) 'Climate Change in the American Mind: October 2015'. Available at: http://climatecommunication.yale.edu/publications/more-Americans-perceive-harm-from-global-warming-survey-finds/, accessed 01/02/17.

Yale (2016a) 'Is There a Climate Spiral of Silence in America?'. Available at: www.climate-communcations.yale.edu/publication/climate-spiral-silence-america, accessed 01/02/17.

Yale (2016b) 'Climate Change in the Mind' www.climatecommuncation.yale.edu/publications/climate-change-American-mind-march-2016/ accessed 01/02/17.

4   The politics of energy

This chapter will focus on energy policy and the energy mix in the US. Energy policy is central to any effective strategy to tackle climate change, specifically a move away from a reliance on fossil fuels such as oil and gas. There will be an analysis of the policy of enhancing the use of so-called renewable energies (such as solar power) as an important step on the road to tackling climate change. However, the chapter will also argue that there is evidence of conflicts and contradictions within US energy policy. The US is caught between the exigencies of tackling climate change and meeting its global obligations such as COP21 and the desire for energy security in an uncertain and changing world. The current natural gas boom in the US brought about by the hydraulic fracking of shale is to be seen in this context. The chapter will analyse the use of hydraulic fracking and its implications for American climate change policy. There will also be an analysis of nuclear power.

Writing in 2010, Abramsky argued that 'the world stands at what is likely to be its last window of opportunity to shift towards a sustainable energy system and avoid the full impact of the crises being fuelled by conventional energy industries' (Abramsky 2010: 78) As we enter 2018, the task of building this sustainable energy system becomes ever more urgent. This is no more so than in the US with its culture of consumerism and excessive use of energy.

The context of energy policy

In 2012, the International Energy Agency (IEA) presented an optimistic outlook as regards energy policy in the US. It argued that the 'global energy map' was being 'redrawn by the resurgence of oil and gas production in the United States' (IEA 2012). Taken together with the IEA's description of the global growth in renewable energy sources such as wind and solar power, the US 'energy economy was becoming – or had already become – an economy transformed' (IEA 2012). It was a different landscape to that which had confronted Barack Obama on his election to the Presidency in November 2008. By 2012 petroleum imports has fallen from 55 per cent of total energy use to 42 per cent. An unprecedented natural gas boom (in large part brought about by the relatively new technique of hydraulic fracking gave the US access to a large new energy source. For Rosenbaum, 'There was even serious discussion of a true American energy independence' (Rosenbaum 2014: 277). The IEA described such a development as 'profound' and stated that 'By around 2020, the United States is projected to become the largest global oil producer' (IEA 2012). Hydraulic fracking was making it possible to extract oil deposits previously considered too technologically challenging and costly. And yet such developments are confronted by one basic truth. For the US to unlock the potential of such a fossil fuel revolution would fly in the face of the evidence of the impact of fossil fuels such as oil and gas on climate change and would weaken international agreements signed up to by the US. I shall return to this shortly.

America's addiction to oil

The US over a long period of time has had a ravenous appetite for energy and the vast bulk of that has been in the form of fossil fuels, in particular oil. The epitome of this is the perception of the US as car culture guzzling up gas (petrol) at an alarming rate. With a few exceptions, major cities in the US have relied heavily over the years on the private car as the major means of transport. This love affair with the car continues to this day. This 'auto-culture', as Wills describes it, has required vast amounts of oil and has resulted in the release of vast quantities of greenhouse gases into the atmosphere (Wills 2013: 145). If the US is to meet its international climate change obligations, this fossil fuel delusion needs to come to an end.

The 1973 Arab states oil embargo in the midst of the Yon Kippur war and the 1975 decision by the oil producing exporting countries (OPEC) to increase by fourfold the price of oil was a missed opportunity by the US (and indeed the broader international community) to take stock of its energy use and needs and to develop an energy strategy for the future not dependent on fossil fuels. Against this background, writing in 1975, Mitchell spoke of 'The energy crisis being a crisis of public policy' (Mitchell 1975: 14). Mitchell was of course writing in a different time. In the mid 1970s the US was heavily reliant on coal and oil. Climate change was not on the policy agenda, and renewable energy was but a small niche market. Yet in many ways Mitchell's description of the public policy challenge of energy is applicable today as it was in 1975. Mancke, writing at about the same time, took a different tack arguing that 'reports of the imminent exhaustion of the Earth's energy resources have been exaggerated'. He went on to say that US energy policy 'should not be concerned with the limits to growth problem at this time' (Mancke 1974: 13). With the benefit of hindsight, such a view can be viewed as naive at best. When we look at the exigencies of US energy policy as we approach the third decade of the twenty-first century, we need to analyse it in the context of the growing threat of climate change and the fact that we live on a planet with finite resources. Such a scenario requires forward thinking and radical solutions.

The environmental impact of oil

In addition to its demonstrable link to climate change, the use of oil has broader environmental consequences. As Rosenbaum has noted, the extraction of oil reserves on federal lands has been a 'source of conflict between proponents of accelerated petroleum production and environmentalism' (Rosenbaum 2014: 81). There are a number of good examples to illustrate this.

The Outer Continental Shelf

First, there is the issue of the conflict over the ecological impact of oil production from submerged public land on the Outer Continental Shelf (OCS), which covers an area of 1.76 billion acres. The OCS is part of the internationally recognised continental shelf of the US which falls under purview of the federal government. It covers a vast area and stretches from Alaska to the coast of the Pacific North West, through the Gulf of Mexico and on to the eastern sea board of the US. The department of industry (DOI) estimates that federal owned OCS lands contain reserves of 8.5 billion barrels of oil and 29.3 trillion cubic feet natural gas (Rosenbaum 2014: 282). At the present time, around 30 per cent of all US domestic petroleum production and 25 per cent of natural gas emanates from the OCS. The environmental lobby has been vociferous in its opposition to further exploitation of the OCS.

Aside from the catastrophic greenhouse gas emissions that could result from the extended exploitation of these fossil fuel resources, there are other more specific environmental concerns. A clear example of this is the BP drilling platform, Deepwater Horizon. Deepwater drilling is a process used in the exploration for oil and gas. There are approximately 3400 deepwater wells in the Gulf of Mexico with depths of more than 150 metres (490 feet). The Deepwater Horizon platform was situated off the coast of Louisiana in the Gulf of Mexico. On 20 April 2010 it exploded and collapsed into the Gulf. The destruction of the platform was the largest accidental marine oil spill in US history, releasing some 4.9 million barrels of crude oil into the Gulf. The result, notes Rosenbaum, was 'an ecological disaster whose long term environmental impact will take decades to assess' (Rosenbaum 2014: 282). The damage to both flora and fauna was extensive. There was extensive damage to marine and wildlife habitats with high death rates reported among dolphins for example. The environmental impact of this ecological disaster is still evident to this day. This major spill brought an understandable reaction from environmental groups who were highly critical of the poor regulatory framework and the inherent risks in exploiting such assets in highly sensitive ecological systems. Looking back on the disaster, the Sierra Club spoke of the Gulf region being used as a 'sacrifice zone' for US energy supplies (Sierra Club 2016).

Investigation by the DOI's inspector general and a bipartisan Congressional committee in 2010 and 2011 were also highly critical of the regulatory oversight by the minerals management service (MMS), the body responsible for regulating oil extraction in the Gulf. Proper oversight could and should have prevented the oil spill. The Secretary of State for the Interior at the time, Kenneth Salazar, promised major reforms at the MMS, describing the findings as 'deeply disturbing' (quoted in Rosenbaum 2014: 282) Echoing my analysis in Chapter 3 of the power of the corporate sector in relation to energy policy and action on climate change, Salazar said the whole affair was 'further evidence of the cosy relationship between some elements of the MMS and the oil and gas industry' (quoted in Rosenbaum 2014: 282).

In September 2014, a US district court judge ruled that BP was primarily responsible for the oil spill due to its gross negligence and reckless conduct (*New York Times*, 4 September 2014). In July 2015, BP agreed to pay $18.7 billion dollars in fines, the largest corporate settlement in US history.

In the aftermath of the Gulf oil spill, controversy and conflict remain the order of the day in relation to the exploitation of oil reserves within the Gulf and the OCS more generally.

The Arctic National Wildlife Refuge

The question of whether to drill for oil in the Arctic National Wildlife Refuge (ANWR) has been an ongoing political and environmental controversy dating backing to the 1970s. The ANWR comprises some 19 million acres of the North Alaskan coast. It is the largest protected wilderness in the US. The protection of the ANWR was enshrined in a 1980 Act of Congress, the Alaska national interest lands conservation act and signed into law by President Carter. Significantly, however, section 1002 of the Act deferred a decision on the management of 1.5 million acres in the coastal plain area of the ANWR. The amount of oil reserves within the region is still a matter for conjecture but studies by the United States Geological Survey (USGS) estimate that there are between 5.7 billion and 16 billion barrels of recoverable crude oil and natural gas within the coastal plain. Oil companies have consistently lobbied to explore and develop these reserves. For those in the environmental movement, the ANWR is considered the jewel in the crown of the Wildlife Refuge System. It is home to some 37 species of land mammal (including caribou and polar bears), eight marine mammals, 42 species of fish and more than 200 migratory birds. A decision to allow extended oil drilling would have major consequences for this delicate and important ecological system.

In January 2015, President Obama announced a proposal to declare an additional five million acres of the ANWR as a wilderness area which would have put a total of 12.8 million acres (including the coastal plain) permanently off limits to drilling and other development. In the same month, in a land mark decision, the US federal fisheries and wildlife service set out the proposals outlined by Obama in a comprehensive conservation plan (CCP). The CCP was signed by the regional director of the Service on 3 April 2015. Obama used his executive authority arising out of the office of the Presidency to back the CCP which was strongly opposed by Alaska Republicans. This was a significant move both by

President Obama (and an important part of his legacy) and of the Fisheries and Wildlife Service, both to protect the environment of the ANWR and as part of a broader strategy to tackle climate change. However, the CCP will require Congressional approval to give it real legislative teeth. With Donald Trump now in the White House, together with a Republican dominated Congress the future of the CCP remains uncertain.

Drilling in the Arctic region

In relation to the broader Arctic region, in November 2016 President Obama (again using his executive authority) reversed his previous support for drilling in the US controlled Arctic region and instead introduced a five year moratorium. In late December of 2016, Obama extended this to a permanent ban. This move has been described as 'a last ditch measure by Obama to strengthen his environmental legacy' (Worland 2016). In making his decision Barack Obama made use of a little known 1953 federal law, the Outer Continental Shelf Lands Act, which empowers the President to exclude US controlled waters form future oil and gas development. Obama's decision was warmly welcomed by environmental groups. For Franz Matzner, director of the National Resource Defense Council's (NRDC) *Beyond Oil Initiative* 'President Obama is stepping outside the normal limited parameters of political calculus (the four year Presidential election cycle)' and instead 'is acting on generational concerns' (Dlouhy 2016). For Matzner, 'In the context of climate change that means taking concrete steps now to trigger results 30 years from now' (Dlouhy 2016).

Indeed this was a major climate change legacy statement by Obama. How it will map out in future policy is a matter for debate. President Obama's successor, Donald Trump, has made clear in the Presidential election campaign and in the early part of his administration of his intention to boost the oil industry by scrapping many Obama era environmental regulations and climate change initiatives. However, given the nature of the 1953 law, it is unclear whether Trump can use his executive authority to overturn the Arctic drilling ban. However, it is likely to be challenged in the courts and the Republican dominated Congress could well reverse the measure.

The North Dakota access pipeline

In Chapter 2 I focused on the controversial Keystone XL pipeline which would bring oil from the tar sands of Alberta, Canada to the US Gulf of Mexico. The pipeline was eventually vetoed by President Obama but the new Trump administration has reinstated it. However, it is likely to be subject to ongoing legal action. No less controversial and politically charged is the case of North Dakota access pipeline. It involves the construction of a 1200 mile pipeline designed to carry oil from North Dakota to a shipping point in Illinois. The cost of the project was estimated at $3.8 billion. The Obama administration came under intense political pressure from local activists and environmental groups to block the project. In December 2016, the Obama administration, under the auspices of the federal government agency, the Army Corps of Engineers, said it would not issue a permit allowing the North Dakota access pipeline to pass under Lake Oahe, a reservoir formed by a dam on the Missouri River. Opponents of the scheme argued that it was a major victory for grass roots activism. For Dallas Goldtooth of the indigenous environmental network 'It was the sheer determination that was shown, the sheer numbers of people who have come to the site'. He went on to argue that activists had shown 'the administration and the oil industry that we are not just a powerless minority, but a powerful majority across the country' (Buncombe 2016).

A key argument in the case against the pipeline was that of environmental justice, an issue that I discussed in Chapter 3. The Standing Rock Sioux, whose reservation lies near the pipeline, opposed the project, arguing that any damage to the pipeline resulting in oil leaks would contaminate their drinking water.

However, the apparent success of local campaigners and environmental groups in blocking the building of the pipeline was dealt a blow when in March of 2017 Donald Trump signed an executive order for the pipeline to go ahead. However, local opposition is set to continue and legal challenges are ongoing. In a further development, Norway's local authority pension fund, KLP, announced that it would sell off shares worth $58 million dollars in companies building the pipeline. The decision came after persistent lobbying by Norway's Sami parliament which represents indigenous people in the country.

The fracking revolution

Kolb argues that geologists have been aware for some time of the considerably large resources of gas and oil held in deep strata of shale and other sedimentary rock formations in certain regions of the US. Such resources were once considered inaccessible due the technological limits. However, the emergence of new technologies such as hydraulic fracking (or fracking as it is more widely known) 'have unlocked these resources and led to an energy renaissance', with the very real possibility that the US could become a net exporter of energy in the not too distant future (Kolb 2014: xvi). There are, however, a number of downsides to this somewhat optimistic scenario, not least the issue of environmental pollution and the impact on climate change. I will return to these shortly.

The process of fracking is highly complex and technical. In broad terms it involves igniting underground explosives to fracture what is known as oil shale. A vertical pipe, often a number of miles deep, is combined with a horizontally drilled pipe which pumps into the shale millions of gallons of salty heated water. This is combined with a variety of chemicals to produce a brine under pressure high enough to penetrate the fractures, and to release the petroleum and natural gas which is embedded in the shale. The mixture is then captured and pumped to the surface.

Frack baby, frack

The US is in the middle of an unprecedented gas and oil drilling rush with the use of fracking to extract both shale gas and oil. Upwards of 19 states possess significant reserves of such gas and oil. In Texas, for example, the Eagle Ford shale geological formation is 400 miles long and 50 miles wide stretching from east Texas to Mexico. It is said to contain one of the largest recoverable oil deposits in the US. In Monterey County California, proposals to frack an area of some 1750 square miles have proved very controversial. Indeed, they have been subject to a legal challenge in respect of the environmental impact. The shale reserves in Monterey contain an estimated 15 billion gallons of oil. The Green River and Baaken shale formations lie under large tracts of Colorado, Montana, North Dakota, Utah and Wyoming. The massive Marcellus shale formation takes in large areas of Alabama, Tennessee, all of Kentucky, West Virginia, the majority of Pennsylvania and a large section of south western New York State.

As Beebeejaun notes, the US has 'become increasingly dependent on the energy' produced by fracking. In the period from 2004 to 2011 there was a 75 per cent increase in its natural gas reserves due to fracking (Beebeejaun 2013). Furthermore as Jenner and Lamadrid note, the US Energy Information Administration (EIA) estimates that the largest five wells in the US, the so-called El Dorados in Pennsylvania, Texas, Louisiana, Arkansas and Michigan, have enough capacity to sustain 100 years of consumption (Jenner and Lamadrid 2013: 442). For the 2016 defeated Democratic Presidential candidate, Hillary Clinton, 'The boom in … natural gas created major economic and strategic opportunities for our country' (Clinton 2014: 522).

Furthermore, fracking is allowing access to oil reserves that were too expensive or too technologically challenging using traditional drilling techniques. The analysts IHS Global Insight have calculated that in 2015 the US produced more oil from fracking than drilling methods. Indeed data from the US EIA shows that in 2010 around one million barrels of shale oil (or tight oil as it is known in the industry). This had risen to two million barrels a day in 2012 followed by a sharp increase to just under five million barrels a day in 2015 (EIA 2016). However, some commentators have raised concerns about the financial prospects for shale/tight oil. In an article in the *Economist* magazine in April 2016, John Castellano of Alix Partners (a debt consultancy firm), writing in the context of the falling price of oil on the global markets, spoke of the threat of bankruptcy of shale producers that had borrowed heavily in 'the boom years'. Indeed two such producers, EnergyXXI and Goodrich Petroleum, filed for Chapter 11 protection (a legal recourse to protect against bankruptcy) in 2016 (Castellano 2016). Further data from the EIA shows that production from shale/tight oil is projected to fall by 700,000 barrels a day in 2017. However, the EIA goes on to argue that production could again start to rise by 2017, dependent on the price of oil on the global markets (EIA 2016).

The fracking revolution has had without doubt a major impact on the energy landscape in the US and presents a major challenge to decades of conventional wisdom about diminishing gas and oil supplies. For Aubrey K McClendon, chairman and chief executive of Chesapeake Energy Corporation, one of the largest shale gas producers in the US, fracking can be viewed as an 'almost divine intervention' (Krause 2008). However, fracking faces strong opposition in a number of quarters with concerns about its environmental impact and the consequences for climate change.

The environmental impact

As we noted above, the fracking process involves the use of millions of gallons of chemically treated water. The return water at the end of the fracking process (known as flowback) contains a mixture of water, chemical toxins and carcinogens which must be purified before the flowback is suitable for other users. Environmental and scientific groups have raised concerns about the ecological and health impact of such a process. In particular, the impact on lakes, rivers and underground water sources has become a matter for debate. As Gabrys notes 'with unconventional (shale) gas and oil, the details of groundwater contamination … become significant energy related problems' (Gabrys 2014: 7). The US Geological Survey has warned that fracking has the potential to contaminate a major underground aqueduct in the Catskill Mountains, a key source of drinking water for New York City (Rosenbaum 2014: 30). The leading US environmental group, the Sierra Club, talks of the 'contamination of surface and ground water' and the huge volumes of toxic water' produced by fracking (Sierra Club 2017). Proponents of fracking (specifically for shale gas) argue that it can play an important role as a transition fuel in the fight against climate change. Hillary Clinton, for example, claims that 'The shift to natural gas is also helping lower carbon emissions because it is cleaner than coal' (Clinton 2014: 522). It is claimed that it produces only 50 per cent of the greenhouse gases produced by conventional fossils fuels such as oil and gas. However, such a figure is open to dispute. Wang *et al.*, for example, make the point that 'serious concerns exist that shale gas generates more greenhouse gases than does coal' (Wang *et al.* 2011: 8196).

The climate change impact of shale oil produced from fracking is even more problematic. The highly respected National Academies of Sciences talks of the concerns about the risks to the environment and human health' in the pursuit of shale gas (National Academies of Sciences 2017). With regard to climate change, it is estimated that shale oil is four times more carbon intensive than oil drilled in the conventional manner.

Rosenbaum has argued that 'Clear and convincing evidence of … environmental impacts attributed to fracking technology … is fragmented and controversial' (Rosenbaum 2014: 4). However, research conducted by the NOAA in conjunction with the University of Colorado estimates that shale gas production in an area known as the Denver Julesburg basin in Colorado is losing 4 per cent of gas (in the form of methane) to the atmosphere. This is worrying as methane is some 25 times more intensive as a greenhouse gas than carbon dioxide

(Tollefson 2012). A 2011 study by Tom Wigley from the National Centre for Atmospheric Research (NCAR) concluded that unless methane leakage rates can be kept below 2 per cent, substituting gas for oil is not an effective means of tackling climate change (NCAR 2011). Indeed, for Kevin Anderson, of the Tyndall Centre for Climate Change 'If we are serious about avoiding serious climate change, the only safe place for shale gas remains in the ground' (quoted in Klein 2014: 214).

Some commentators such as Professor William Press, a member of the President's Council of Advisers on Science and Technology, still argue that the climate change goals outlined by President Obama can be achieved by supporting further wide scale fracking over the next few years (quoted in the *Observer*, 17 February 2013). But there is a body of scientific evidence highlighting the impact of fracking on the environment and attempts to tackle climate change.

The EPA has been conducting its own research into the potential negative impacts on the environment and human health of fracking. Its report was published in December 2016 (EPA 2016). The EPA confirmed that hydraulic fracking has a negative impact on the quality of drinking water in certain circumstances. The EPA report lists a number of concerns. These include:

- The withdrawal of water for hydraulic fracking particularly in areas with limited or declining groundwater resources.
- The injection of hydraulic fracking fluids into wells with inadequate mechanical integrity, thereby allowing gases or liquids to move into ground water resources.
- The discharge of inadequately treated hydraulic fracking waste water into surface water.

The hydraulic fracking boom in the US has certainly contributed to its energy security but the debate about its impact on climate change and the broader environment is certain to continue for some time to come.

The demise of King Coal

For Rosenbaum, American presidents from Richard Nixon to George W. Bush have 'tried in one way or another to dam the flow of imported oil' into the US with a 'wall of coal' (Rosenbaum 2014: 287). This was in large part a strategy to provide greater energy security and to reduce its reliance on oil from the increasingly unstable Middle East. It is estimated that at current levels of consumption, there are some 250 years of coal reserves in the US. As a consequence argues Rosenbaum it was not surprising that 'federal energy planners should repeatedly attempt to substitute abundant domestic coal for expensive, insecure, imported oil' (Rosenbaum 2014: 287). However, the exigencies of climate change would seem to render coal as irrelevant as a long term energy source for the US. However, as in all such matters politics comes to the fore. President Obama early on in his administration outlined his plans to tackle climate change. Yet, as

Rosenbaum notes, he had to walk a 'political tightrope through the turbulence of the coal controversies' (Rosenbaum 2014: 287), with his home state of Illinois being a significant area for coal production.

The emergence of hydraulic fracking has seen natural or shale gas emerge as a key energy source in the US and pose a major threat to the coal industry in the US. Indeed as Levi observes, 'There is little question that the first few years of the shale boom have slammed coal' (Levi 2014: 98). Shale gas still contributes to climate change but produces only 50 per cent of the carbon emissions compared to coal. It presents a major threat to coal jobs in a number of US states such as Illinois, Pennsylvania and West Virginia. In addition the CPP (see Chapter 2) would most certainly lead to the closure of a number of coal energy plants. The CPP became a difficult political sell in some quarters. Donald Trump made political capital during the 2016 Presidential election promising to restore coal to its former glory as part of his pledge 'to make America great again'. Trump's strategy bore some electoral success in the 'rust belt' with surprise wins in the traditionally Democratic coal state of Pennsylvania, as well Michigan and Wisconsin.

The coal industry has been in long term decline in the US, as is in the case in many countries across the globe. In 1980 there were in the region of 230,000 coal mining jobs. In 2016 that figure had shrunk to less than 80,000. There are a number of key questions to be asked about Trump's coal plan. How easy will it be to get the private sector to put in investment to access US coal reserves? As we have seen above hydraulic fracking for both shale gas and oil would seem a far more lucrative pursuit for the private sector than coal. Unless of course President Trump intends to revive the American coal industry through large federal government subsidies, or even a government run coal industry. This seems very unlikely given that US political culture runs against big government.

Trump's commitment to put coal at the centre of his energy strategy (if we can call it that) would drive a coach and horses through a number of global agreements on climate change which the US has signed up to. However, this is of a piece with Trump's views on climate change as part of a Chinese conspiracy to undermine the US economy. He also made a number of statements that he would withdraw the US from COP21. How this will pan out in substantive policy terms time alone will tell but it certainly marks a major change in tone from the Obama years.

There is however a view in some circles that the US could still have a viable coal industry and still keep its global commitments to tackle climate change. It is argued that the solution lies in the so-called clean coal technology of carbon capture and storage (CCS). CCS can be described as a somewhat experimental and risky technology. It attempts to capture carbon dioxide from fossil fuels such as coal used in industrial process (for example factories and energy power plants) and store it underground in saline aquifers or redundant oil wells, thus preventing its release into the atmosphere. However, at the moment, there exists no such commercial technology on any appropriate scale. For Robert Engelman (2012), President of the Worldwatch Foundation 'CCS is worth exploring as one of a large array of potential strategies for slowing the build up of CO_2 in the atmosphere'. However, to achieve this requires significant levels of investment, both public and private.

As part of an attempt to develop clean coal technology, under the Obama Presidency, the 2009 American recovery and reinvestment act was passed. It included a $775 billion investment package, the aim of which was to jump start the American economy in the wake of the 2007/2008 global financial crisis. It included a package of some $3.4 billion for research and development into CCS. The Centre for American Progress has argued that 'underground storage capacity in the United States is believed to be ample and widespread, and long term leakage of CO_2 from properly permitted and monitored storage reservoirs is expected to be negligible' (Centre for American Progress 2011) In 2010, with significant financial backing from the Department of Energy, the American Electric Power Company launched one of the earliest CCS pilot projects at its Mountaineer power plant in New Haven, West Virginia. However, English Electric cancelled the project in 2011. This came as a big blow as significant political capital had been put into this project. Levi refers to the 'scepticism about CCS', pointing out that no full scale power plant using CCS has ever being built. He argues that the costs of a coal fired power plant using CCS 'would inevitably cost north of a billion dollars to build' (Levi 2014: 171). Thus the prospect for the future funding of CCS remains very uncertain to put it mildly.

The promised land of nuclear power

H. G. Wells' 1914 novel *The World Set Free* tells the prophetic story of man's harnessing of the (at that time) newly discovered power of the atom and how this power nearly destroys civilisation in a catastrophic war. Others have viewed the atom as having a more peaceful purpose, producing an endless stream of low cost power. For example, on 8 December 1953, US President Dwight D. Eisenhower, delivered a speech to UN general assembly entitled 'Atoms for Peace'. In his speech he presented the case for nuclear power as an important energy source. This was in the context of an emerging cold war between the US and the Soviet Union, and an American public increasingly concerned about the threat of nuclear arms.

More recently, a number of academic commentators and indeed some prominent environmental activists have extolled the virtue of nuclear power as an important energy in the fight against climate change. Nuclear power is a carbon-free technology. As such it does not produce greenhouse gases. James Lovelock, a leading environmentalist and author of the Gaia Hypothesis, has argued that climate change is such a major threat to the planet that we need to make full use of nuclear power. For Lovelock, now is not the time to experiment with 'visionary energy solutions' (Lovelock 2004). In the US, Stewart Brand, an important figure in the modern environmental movement and the editor of *Whole Earth Catalogue* has argued that 'nuclear is green' (Brand 2012). Another key figure in the environmental movement and author of *Six degrees; our future on a hotter planet*, Mark Lynas argues for the wider use of nuclear power 'in order to avoid more carbon emissions' (Lynas 2012).

However, an event on 28 March 1979 shook the nuclear industry in the US both literally and metaphorically. The event in question involved the Three Mile Island nuclear plant, located in the middle of the Susquehanna River, ten miles from Harrisburg, Pennsylvania. A technical fault combined with human error led to a meltdown of 45 per cent of the core in the plant's nuclear reactor which resulted in high levels of radioactive fallout in the area. It led to widespread panic in the local community around the plant. As Graetz notes:

> The Three Mile Island accident and the panic that it engendered marked the end of the widespread hopes and expectations of the 1960s and early 1970s that nuclear power would soon supply the major share of the nation's economy.
>
> (Graetz 2011: 63)

The incident at Three Mile Island raised major public concerns about the safety of nuclear power, both locally and across the US and led to well organised opposition to the building of future nuclear plants from environmental and citizen groups. In 2015, the US produced 20 per cent of its electricity from 99 plants across 30 states operated by 30 different power companies. This was 'far less than was expected and hoped for' at the beginning of the 1970s (Graetz 2011: 77).

Mark Lynas, whilst being an advocate of nuclear power as an expedient way to deal with the challenge of climate change in the short term, does acknowledge its risks, specifically with regard to the environmental impact of nuclear waste and the danger of nuclear accidents (Lynas 2008: 273). Such risks give rise to concerns amongst policy makers and politicians to this day. For example in 2010, after an intervention by President Obama, federal funding for a permanent nuclear waste repository at Yucca Mountain, 100 miles from Las Vegas, was brought to an end. This left no designated long term storage site for nuclear waste in the US. As a consequence most nuclear waste, including highly radio-active fuel rods, is stored on site in steel or concrete casks, itself a significant environmental risk.

Some of the nuclear power stations dating back to the 1970s are showing signs of wear and tear. For example in May 2009, there was a discovery of a 100,000 gallon leak of coolant waste from an underground pipe at the 40 year old Indian Point 2 nuclear in Buchanan, New York State. The plant sits on the Hudson River, outside New York City. The leak raised serious concerns about its impact on public health and the environment. In 2016, Indian Point 2 encountered further problems as it was forced to close to repair another water leak into the Hudson River. Over the last five years there have been recorded leaks and other safety incidents at a number of other nuclear power facilities including three plants in Illinois, one in New Mexico and one in Florida.

The context here is that a number of these ageing plants have been seeking to have their operating licences renewed, raising further long-term safety concerns.

In the global context, the March 2011 Fukushima nuclear accident in Japan raised concerns amongst policy makers and the public about the safety and efficacy of nuclear power.

Rosenbaum has argued that 'The halt in developing nuclear power stations struck a serious blow' at the US 'efforts to achieve energy independence with a carbon free fuel' (Rosenbaum 2014: 77). What Rosenbaum is implying is that as a non-fossil energy source, nuclear power can make an important contribution to the fight against climate change as it does not produce carbon dioxide, one of the key greenhouse gases that cause climate change.

However, other commentators question the idea of nuclear power as a non-carbon fuel source. Caldicott, for example, argues that 'Nuclear energy creates significant greenhouse gases and pollution' (Caldicott 2006: vii). The reason for this argues Caldicott is that large quantities of fossil fuels are needed to mine and refine the uranium which is required to make nuclear power stations function and to build the large infrastructure required for nuclear power stations. (Caldicott 2006: vii). The use of such fossil fuels emits large quantities of carbon dioxide into the atmosphere.

There are other inhibiters to the use of nuclear power. Caldicott talks of the huge costs of nuclear power, what she describes as 'the disastrous economics of nuclear power' (Caldicott 2006: 163). As a consequence, any order for a new nuclear reactor would require a large subsidy from the public purse. This is money that would be better spent, argues Caldicott, on more sustainable energy resources such as renewables (Caldicott 2006: 163). It is to this that I now turn.

The renewables revolution

For those seeking to combat climate change, a key part of the solution lies in renewable energy technologies. Indeed the latest world energy outlook from the IEA states that due to major transitions in the global energy system over the next decades (a product of COP21 and other actions on climate change), renewables (along with natural gas) are the biggest winners in the race to meet energy demand by 2040 (IEA 2016). In broad terms, one can define renewable energy as energy that is produced from resources that do not deplete when their energy is harnessed. Such resources include solar, wind and wave power. This is in contrast to fossil fuels such oil, coal and gas which are finite. In contrast to fossil fuels, renewable energy sources have a low carbon impact.

For the US Department of Energy (DOE), 'A clean energy revolution is taking place across America, underscored by the steady expansion of the US renewable energy sector' (DOE 2017). In her memoir, the former Secretary of State, Hillary Clinton, writes of how the Obama administration promoted clean energy sources such as solar, wind and hydro electric (Clinton 2014: 521). However, Mitchell writes of how the US is moving 'with a dangerous slowness towards the use of renewable sources of energy' (Mitchell 2013: 266). Is this an accurate analysis? As we can see in Table 4.1, renewables only add up to 10 per cent of all energy consumption in the US.

Table 4.1 US energy use by source 2015 as a percentage

Petrol	36%
Natural gas	29%
Coal	16%
Renewables	10%
Nuclear	9%

Source: US Energy Information Service (eia.org).

In Table 4.2 we can see the breakdown of how this 10 per cent is spread across different renewable energy sources.

If the US is to meet its international obligations on climate change, its percentage of renewable energy needs to increase significantly. Swisher and Porter writing in 2006 argued that 'policy support must be consistent, predictable and long term if renewable energy is going to make a significant contribution' to American energy use (Swisher and Porter 2006: 185). The Obama administration invested a great deal of political capital in renewable technology. Obama took office in January 2009 to a back drop of global financial crisis. One of the first significant developments of the Obama Presidency was the passing of the American Recovery and Investment Act (ARRA). The Act set aside some $80billion dollars for the development of clean energy technologies and jobs. This included loans and subsidies for the manufacture and commercial distribution of advanced solar and wind technology. Obama had himself wanted to go further but the final package was a compromise after much political bargaining with Congress.

On Earth Day, 22 April 2009, in an attempt to show his green energy credentials, Obama visited Trinity Structural Towers in Iowa, a plant that manufactured wind energy equipment. 'The choice we face' the President said 'is not between saving our environment and saving our economy. The choice we face is between prosperity and decline. We can remain the world's leading importer of oil, or we can become the world's leading exporter of clean energy' (Graetz 2011: 168).

Renewable energy sources were an important element of Obama's climate action plan (see Chapter 2). However, Obama's support for renewable energy did contain some ambiguities. For example in a speech to students at Georgetown University

Table 4.2 US renewable energy source 2015 as a percentage

Hydroelectric	25%
Biofuels	22%
Wood	21%
Wind	19%
Solar	6%
Biomass waste	5%
Geothermal	2%

Source: US Energy Information Service Centre (eia.org).

on 22 June 2013, Obama outlined his strategy to tackle climate change including the important role of boosting renewable energy. However, in the same speech, he welcomed the US shale gas boom as an important transitional fuel. Obama's support here for shale gas can be seen in the context of the need to meet the exigency of energy security and to make the US less dependent on foreign oil, in particular from unstable areas of the Middle East. However, renewables also have the potential to address concerns over energy security by reducing the dependency of US on foreign imports.

Commentators such as Rosenbaum acknowledge the significantly increased commitment to investment in renewable energy sources in the US over the last decade. However, he goes on to argue that 'fossil fuels are certain to remain the nation's primary energy resource well beyond 2050' (Rosenbaum 2014: 310). Indeed, as we have noted, renewable energy use still only makes up just 10 per cent of energy use in the US (see Table 4.1); however, it is a sector that is growing. At the end of 2016, the US had 40 gigawatts (GW) of installed photovoltaic capacity, almost double that for 2015. There are approximately 6000 solar projects under way. For example, in the summer of 2013 the Chinese company Trina Solar won a contract to supply one million solar photovoltaic panels for a 250 megawatt power facility in the Nevada desert. In California, there is the 250 megawatt valley solar ranch. Based in San Luis Obispo County, the site was built by NRG Energy Inc and SunPower Corporation. It powers some 42,000 homes. In addition, wind power has seen a growth and now accounts for around 19 per cent of all renewable energy.

The second largest source of renewable energy in the US is biofuels (at 22 per cent). The principle biofuel in the US is ethanol. Ethanol is a transportation fuel which is mainly distilled from corn (it can also come from grains such as barley and sorghum. It is usually blended with petroleum for use in motor vehicles. Nearly all gasoline (petrol) in the US is 10 per cent by volume ethanol. Proponents of biofuels such as ethanol argue that it can make an important contribution to the fight against climate change as it has a neutral impact in terms of carbon dioxide. However, environmental campaign groups cast major doubt on this. They point to the large-scale deforestation that is taking place to provide the land necessary to grow crops such as a corn (a central element in ethanol). Such deforestation leads to the loss of important carbon sinks (trees soak up carbon gases such as carbon dioxide) and thus contributes to climate change (Campaign against Climate Change 2017). The use of ethanol has other implications. In the US, farmers have seen the potential increase in profits that can be made from growing corn for ethanol (as opposed for food). The resultant switch in production has 'resulted in a significant increase in food prices worldwide' (Cahill 2011: 239). Such production has also led to a series of food riots in developing countries over the last few years. This led to the then UN special rapporteur on the right to food, Jean Ziegler, to declare in 2013 that biofuels such as ethanol were 'a crime against humanity' (as quoted in the *Guardian*, 26 November 2013).

Jobs and the green economy

Proponents of renewable or clean energy argue that their use is not only benefi-
cial for the environment and the fight against climate change but can also result
in a 'Green Economy' which can create high skill green jobs. However, commentators such as Graetz are sceptical of the potential of such green jobs. The
US lost some 5.6 million manufacturing jobs in the period 2000 to 2010 alone.
The so-called rust belt in the American mid west with its decaying factories and
redundant steel works is the physical embodiment of this industrial decline. 'If
the idea is that green energy jobs' writes Graetz 'will make up for the manufacturing jobs we have lost, it seems Pollyannaish' (Graetz 2011: 168).

Whilst there may be validity in Graetz's line of argument, there is evidence of
growing employment in the renewable energy sector. A recently published report
by the DOE shows that there are in the region of 300,000 jobs in solar energy
generation as of the end of 2015, with a 28 per cent increase from to 2014 to
2015 (DOE 2017: 28). Solar panel installation companies employ 57 per cent of
those working in the solar sector. Since 2010, solar installation firms have seen
employment rise from 44,000 to 120,000. That is an increase of some 173 per
cent. Combining those working in the wind power sector (standing at some
77,000) this is gives us a total of some 377,000 solar and wind power jobs. This
compares favourably to the 93,000 people employed in the coal industry.
However, it is dwarfed by the 9.8 million jobs in the oil and gas industry (DOE,
2017). For David Foster, senior advisor on industrial and economic policy at the
DOE, the DOE report 'verifies the dynamic role our energy technologies play in
21st century economy' (Bulman 2017).

Conclusion

The above discussion on renewable energy leads on to a fundamental question of
the kind of radical change needed to tackle climate change and promote sustainability in the US more generally. Roberts has spoken of what he describes of the
'myth of the perfect gadget', this being the idea that 'solving the global energy
problem is only a matter of waiting for the right technology' (Roberts 2005:
260). Of course, technology such as solar and wind power has an important role
to play in tackling climate change, but it has to be seen in the context of broader
social, economic and political change. 'A common assumption' notes Jones 'is
that we (the USA) can simply replace cold fired power stations with wind tur-
bines and expect everything else to remain the same' (Jones 2014: 1). But for
Jones this is simply incorrect. Rather 'we need to incorporate broader range of
perspectives' (Jones 2014: 5). This could include less of a focus on consumerism
and buying less stuff, more localised production, putting human well being
before economic growth, reduced use of the car and re engaging with nature so
that it has value in itself and is not merely a resource for human exploitation.
This is a major challenge (especially so in a country like the US) and requires us
to look at the world through a different lens. Both domestically in the US, and

on the wider international scene, we need a radical rethink of how we do politics and economics. This requires innovation and dynamism.

References

Abramsky, K. (2010) *Sparking a Worldwide Energy Revolution: Social Struggles in the Transition to a Post Petrol World*, Oakland, CA: AK Press.

Beebeejaun, Y. (2013) 'The Politics of Fracking: a Public Policy Dilemma' *Political Insight*, vol. 4, issue 3, pp. 18–21.

Brand, S. (2012) 'Some Environmentalists Back Nuclear Power' *San Francisco Chronicle*, 13 June.

Bulman, M. (2017) 'US Solar Power Employs More People than Oil, Coal and Gas Combined, Report Shows' *Independent*, 24 January.

Buncombe, A. (2016) 'Sioux Defeat Plan for Oil Pipeline near Reservation' *The I*, 6 December.

Cahill, M. (2011) 'Transport' in Fitzpatrick, T. (ed.), *Understanding the Environment and Social Policy*. Bristol: Policy Press.

Caldicott, H. (2006) *Nuclear Power is Not the Answer*. New York: New Press.

Castellano, J. (2016) 'Oil Markets, Drill Will' *The Economist*, 23–29 April.

Clinton, H. (2014) *Hard Choices*. New York: Simon and Schuster.

EPA (2016) 'Hydraulic Fracturing for Oil and Gas: Impact from the Hydraulic Fracturing Water Cycle on Drinking Water Resources in the United States, Final Report' Washington, DC: EPA.

Gabrys, J. (2014) 'Powering the Digital: From Energy Ecologies to Electronic Environmentalism' in Maxwell, R., Raundalen, J. and Vestbergy, N. (eds), *Media and the Ecological Crisis*, Abingdon, Oxon: Routledge.

Graetz, M. (2011) *The End of Energy: The Unmaking of American Environmental Security and Independence*, Cambridge, MA: MIT Press.

IEA (2012) *World Energy Outlook 2012: Executive Summary*, Paris: France.

IEA (2016) *World Energy Outlook 2016: Executive Summary*, Paris: France.

Jones, C. (2014) *Routes of Power: Energy and Modern America*, Cambridge, MA: Harvard University Press.

Klein, N. (2014) *This Changes Everything: Capitalism vs. the Climate*, London: Allen & Unwin.

Kolb, R. (2014) *The Natural Gas Revolution: At the Pivot of the World's Energy Future*, Upper Saddle River, NJ: Pearson.

Krauss, C. (2008) 'Drilling Boom Revives Hopes for Natural Gas' *New York Times*, 25 August.

Jenner, S. and Lamadrid, A. (2013) 'Shale Gas vs. Coal: Policy Implications from Environmental Impact Comparisons of Shale Gas, Conventional Gas, and Coal on Air, Water, and Land in the United States' *Energy Policy*, vol. 53, pp. 442–453.

Levi, M. (2014) *The Power Surge: Energy, Opportunity, and the Battle for America's Future*, Oxford: Oxford University Press.

Lovelock, J. (2004) 'Nuclear power is the only green solution' *Independent*, 24 May.

Lynas, M. (2008) *Six Degrees: Our Future on a Hotter Planet*, London: Harper Perrenial.

Mancke, R. (1974) *The Failure of US Energy Policy*, New York: Colombia University Press.

Mitchell, E. (1975) *US Energy Policy: A Primer*, Washington, DC: American Institute for Public Policy Research.

Mitchell, T. (2013) *Carbon Democracy: Political Power in the Age of Oil*, London: Verso.

Roberts, P. (2005) *The End of Oil: On the Edge of a Perilous World*, New York: Mariner Books.

Rosenbaum, W. (2014) *Environmental Politics and Policy*, London: Sage.

Swisher, R. and Porter, K. (2006) 'Renewable Policy Lessons from the US: The Need for Consistent and Stable Policies' in Mallon, K. (ed.), *Renewables, Energy Policy and Politics: A Handbook for Decision Making*, Abingdon, Oxon: Routledge Earthscan.

Tollefson, J. (2012) 'Air Sampling Reveals High Emissions from Gas Field: Methane Leaks During Production may Offset Climate Benefits of Natural Gas' *Nature*, vol. 482, issue 7384, pp. 139–140.

Wang, J., Ryan, D. and Anthony, E. (2011) 'Reducing the Greenhouse Gas Footprint of Shale Gas' *Energy Policy*, vol. 39, issue 12, pp. 8196–8199.

Wills, J. (2013) *US Environmental History: Inviting Doomsday*, Edinburgh: Edinburgh University Press.

Worland, J. (2016) 'President Obama Issues Permanent Arctic Drilling Ban' *Time Magazine*, 20 December.

Internet sources

Campaign against Climate Change (2017) 'Bio Fuels and Climate Change: The Facts'. Available at: www.campaigncc.org/biofuels, accessed 31/01/17.

Centre for American Progress (2011) 'Coal Capture and Sequestration'. Available at: www.Americanprogress.org/issues/2009/03ccs_101.html, accessed at 13/06/11.

Dlouhy, J. (2016) 'Offshore Drilling Foes Invoke 1953 Law to Prod Obama on US 2011 Ban'. Available at: www.bloomberg.com/news/articles/2016-05-24/offshore-drilling-foes-invoke-1953-law-to-prod-obama-on-u-s-ban, accessed 31/01/17.

DOE (2017) 'A Clean Energy Revolution is Taking Place Across America'. Available at: www.energy.gov/science-innovation/energy-sources/renewable-energy accessed 21/03/17.

EIA (2016) 'Future US Tight Oil and Shale Gas Production Depend on Resources, Technology and Markets'. Available at: www.eia.gov/todayinenergy/detail.php?id=27612, accessed 02/02/17.

Engelman, R. (2012) 'Growth of Carbon Capture Stalled in 2011'. Available at: www.worldwatch.org/growth-carbon-capture-and-storage-stalled-2011, accessed 02/02/14.

National Academies of Sciences (2017) Available at: http://sites.nationalacademies.org/DBASSE/BECS/DBASSE_083187, accessed 01/05/17.

NCAR (2011) 'Switching from coal to natural gas would do little for global climate, study indicates'. Available at: www2.ucar.edu/atmosnews/news/5292/switching-coal-natural-gas-would-do-little-global-climate-study-indicates, accessed 04/09/14.

Sierra Club (2016) 'Learning our Lessons from Deep Water Horizon'. Available at: https://ventana2.sierraclub.org/lay-of-the-land/2016/04/learning-our-lesson-deepwater-horizon, accessed 31/03/17.

Sierra Club (2017) Available at: www.sierraclub.org/new-jersey/stop-fracking, accessed 24/01/17.

5 Action on climate change at state and city level

Introduction

This chapter will argue that the relative lack of action on climate change at the federal level of government in the US has created a policy space and a regulatory vacuum which a number of US states and cities have sought to fill in a substantive way, creating an arena of vibrant debate and activity. While the Bush administration rejected the Kyoto Protocol, many states and cities responded in a different way, adapting and implementing a number of the Kyoto principles in their climate change action policies. There will be an analysis of the policy initiatives of a number of these states and cities.

Action at the state level

As Brewer notes, 'sub national government entities have much leeway in the context of the making and shaping of policy in the US federal system of government' (Brewer 2014: 117). A number of state governments have taken advantage of this leeway to develop strategies to tackle climate change in the context of federal inertia, in particular during the two terms Presidency of George W. Bush. As Rabe notes, 'many states launched unilateral policy experiments' (Rabe 2015: 63). A particular focus for a number of states was renewable energy and energy efficiency. However, state action on climate change has faced a number of challenges in recent years. For example, it is important to note that there is a great deal of variation between states on the climate change policy agenda, together with both inter and intra party tension and conflict. In addition, the climate change policies of a number of states have led to a political backlash with resultant changes in political control from Democrat to Republican at the state level. This has resulted in policy retrenchment and policy reversal in a number of cases. Despite this Sussman and Daynes stress the 'important role of the states in addressing climate change, given inaction by the federal government' (Sussman and Daynes 2013: 4).

State action on climate change networks and co-operation

Over the past decade, there have been a number of attempts to forge networks and co-operation across states which have met with varying degrees of success.

One such example is the western climate initiative (WCI). The WCI was broadly based on Kyoto principles. Set up in 2007, the WCI was a partnership between seven US states and four Canadian provinces which sought to set up a cap and trade regime with the goal of a 15 per cent reduction in greenhouse gas emissions from 2005 levels by 2020. The seven US states were California, New Mexico, Arizona, Washington, Oregon, Montana and Utah. However, swings in political control from Democrat to Republican in states such as Arizona, New Mexico and Utah proved a significant factor in the collapse of the WCI. This left only California and four Canadian provinces, British Columbia, Manitoba, Ontario and Quebec. By 2013 only California and the Canadian province of Quebec remained. However, despite such setbacks, certain states of the former WCI still continue to pursue climate change initiatives on an individual level. I will return to this shortly.

Another example of a network state response to tackling climate change is that of the regional greenhouse gas initiative (RGGI) which was signed by seven north eastern states of the US in December 2005. The seven states were Connecticut, Delaware, Maine, New Hampshire, New York, New Jersey and Vermont. It was the first market based regulatory programme in the US to reduce greenhouse gas emissions involving a cap and trade system. It set out mandatory targets for greenhouse gas emissions in the electricity generating sector with a start date of 2009. The agreement focused on fossil fuel plants with 25 MW or greater generating capacity. Its aim was to stabilise greenhouse gas levels through to 2015 together with a target of reducing such greenhouse gas emissions by 10 per cent by 2019. The initiative has faced many challenges, not the least of which was the decision of the state of New Jersey governor, the Republican Chris Christie, to leave the scheme in May 2011. Yet as Rabe has noted, the RGGI has kept up quarterly carbon allowances (a central feature of the cap and trade scheme (Rabe 2015: 64). This has generated significant funds for state renewable energy and energy efficiency programmes.

One final example is that of the mid western greenhouse gas accord which was set in place with the signing of a memorandum of understanding in November 2007. The signatories were the US states of Illinois, Iowa, Kansas, Michigan, Minnesota and Wisconsin together with the Canadian province of Manitoba. Its aim was to introduce greenhouse gas reduction targets based on a market based cap and trade system. Rabe notes the very limited impact of the accord, arguing that the memorandum is essentially defunct, having been effectively ignored by all the participants (Rabe 2015: 64). Such a limited impact needs to be seen in the context of the failure at the federal level of government in Washington to pass a cap and trade bill (see Chapter 2). The accord has been inactive since March 2010.

Policy action at the state level

At the individual state level there have been a number of substantive policy initiatives to tackle climate change and it is to this that I now turn. It should be noted that Democratic leaning states are more likely to have in place substantive policies to tackle climate change than Republican leaning states.

One policy approach to tackling climate change which has gained political traction in a number of US states is what is known as an energy efficiency resource standard (EERS). It is a performance based mechanism that requires electricity and natural gas distributors to achieve a percentage of energy efficiency relative to a baseline. Some 25 states have now established some form of EERS (from baseline energy efficiency measures to support for renewables). According to the American Council for an Energy Efficient Economy (ACEEE) a number of states have developed a range of other energy efficiency initiatives such as building codes and the purchase of high efficiency vehicles and light. The ACEEE's 2016 edition of the state energy efficiency scorecard ranks all 50 states and the District of Columbia. California and Massachusetts tie for first place with Vermont, Rhode Island and New York rounding off the top five, all Democratic states (ACEEE 2016). Energy efficiency measures add up to a not inconsiderable contribution to tackling climate change. As Linda Breggin of the Environmental Law Institute has observed, 'What is remarkable is not that California is leading the country but how many other states are on the move as well' (Breggin 2012).

But what of broader action by US states on climate change? Recent analysis by the Center for Climate and Energy Solutions (C2ES) shows that 34 states have in place a climate action plan (CAP), setting out targets to reduce greenhouse gas emissions with concomitant policy actions. The majority of these states are on the western seaboard, the mid west and the east coast, and are predominantly Democrat leaning states. However, based on the research of the C2ES, a number of Republican states are listed as having a CAP. They are Arkansas, the Carolinas, Georgia, Kentucky, Montana and Utah (C2ES 2016).

An analysis of the 50 states of American shows a variety of practice and action on climate and environmental policy. Here is a snap shot to illustrate this. Two states, Maryland and New York (both Democrat) have in place CAPs and have also have voted to ban hydraulic fracking. The state of Alabama (Republican) does not have a CAP in place but has implemented in house energy efficiency measures. An analysis of the web site of the state of Louisiana (Republican) does show some discussion on the causes of climate change. There are also a number of policy documents on the role of renewable energy. However, there is no CAP and no targets set for the reduction of greenhouse gases. The state of Minnesota, with a Democrat governor Mark Brandt, not only has a CAP but also sent delegates to the UN Paris COP21 conference in December 2015. The traditionally Democratic state of Washington in the Pacific North West also has its own CAP. It is also known as the evergreen state as three quarters of its electricity comes from hydroelectric power. On 8 November 2016, after a grass roots campaign, Initiative 732 made it on to the ballot. If successful it would have been the first time that a carbon tax had been implemented in the US (it was based on a similar scheme in British Columbia, Canada). Carbon emissions were to be taxed at a rate reaching $25 per tonne by 2018, with a subsequent annual rise of 3.5 per cent. The proposal was spearheaded by the environmental economist, Yoram Bauman. However, initiative 732 was subsequently rejected. However, the fact that such a proposal got on to the ballot in the first place was significant (*Economist* 2016).

Having looked at a number of vignettes of state activity around climate and environmental policy, I will now look at the climate change and environmental policies of a number of selected states in more detail.

California

As Brewer notes, California is one of the most advanced states in its response to climate change (Brewer 2014: 117). This is significant as one in eight Americans live in the state of California and it is also the sixth largest economy in the world. In 2002, California became the first US state to compel the automobile industry to start reducing emissions of CO_2 as part of a strategy to improve air quality. In 2006 the California state legislature passed into the law the California Global Warming Solutions Act. The Act forms the basis of California's CAP and gives the state significant powers to shape policy, draw up regulations and allow enforcement to reduce greenhouse gases such as CO_2. The Act has been highly contested and has been the subject of much lobbying, notably from the automobile industry with legal redress sought in the courts Nonetheless, it is a significant piece of legislation, not least because it was passed with the support of the then Republican governor of California, Arnold Schwarzenegger. It caps greenhouse gas emissions in key sectors of the Californian economy (covering in the region of 85 per cent of the state's greenhouse emissions) with a target to cut greenhouse gases by 15 per cent below 2005 levels by 2020 and 80 per cent below 1990 levels by 2050. This is in line with the target set the by the UN IPCC. It is based on a market based cap and trade system. As Rabe notes 'California now represents the most prominent North American experiment attempting cap and trade' (Rabe 2015: 66). More recently, in August 2016, the California state legislature passed a bill to cut greenhouse gases by 40 per cent from 1990 levels by 2030. The bill also restated the target of an 80 per cent cut by 2050.

Looking back on his term as Governor, Arnold Schwarzenegger claimed California was at the forefront at the global level in shaping policies to tackle climate change. Posting on his Facebook page on 15 December 2015, the former governor argued that:

> Renewable energy is great for the economy, and you don't have to take my word for it. California has some of the most revolutionary environmental laws in the United States. We get 40% of our power from renewables and we are 40% more energy efficient than the rest of the country.

In addition, California has put in place regulations to reduce greenhouse gas emissions from automobiles using the Clean Air Act as the legal underpinning for this. This matter has been the subject of much legal wrangling and political conflict but in June 2009 the EPA, following an intervention from President Obama, granted California a so-called 'waiver' thus enabling the state to enforce its greenhouse gas emission standards for new vehicles. The state of California has also been active in the area of renewable energy. Legislation brought into law in April 2011 increased its targets for renewables for 2020 from 20 to 33 per cent.

While it is true to say that California's policies to tackle climate change enjoy a significant degree of political support, the overall picture is complex. The battle to tackle climate change in California has brought to the fore the complex interplay and conflict between a variety of interest groups. The environmental policy arena is a crowded one with many actors and groups seeking to advance their interests or indeed to block the interests of others. Propositions (similar to referendums) are a common feature in California as they are in a number of US states. In November 2006 Proposition 87, backed by a number of environmental groups, called for the setting up of a fund to support renewable forms of energy. The proposition, if successful, would have established $4 billion programme to reduce petroleum consumption by 25 per cent. It would also have introduced incentives for alternative energy and other energy efficiency technologies. It was to be funded by a tax on the producers of oil extracted in California. However, it met with intensive lobbying by the oil industry and was subsequently defeated. The No campaign was funded in large part by the Chevron Corporation and Aera Energy. In November 2010, Californians voted on Proposition 23. It was backed by a number of corporations which included Valero Service, an oil production and distribution company, Tesoro Companies, refiners of petroleum products, and Marathon Petroleum Company. If passed, proposition 23 would have effectively put paid to the Global Warming Solutions Act. In the event, it was decisively defeated by a margin of 61 to 30 per cent. Environmental lobby groups played an important result in the outcome. They included the National Wildlife Federation, the League of Conservation Voters, the National Conservancy and the Sierra Club.

California was an early pioneer in policy initiatives to tackle climate change, both in the US and on a global level. Speaking at the Democratic Presidential Convention on 28 July 2016, the Democrat Governor of California, talked of how climate change was 'the existential threat of our time'. He spoke of how California had some of 'the toughest climate laws in the country'. Yet as Rabe has noted, 'California political and climate policy leaders would no doubt welcome' more support from other states in the Union (Rabe 2015: 67). I will now focus on climate change policy initiatives in a number of other US states.

Illinois

The state of Illinois has been one of the more active states in setting out a strategy to tackle climate change under the leadership of Pat Quinn who was Democratic Governor of the State from January 2009 to January 2015. A former official I spoke to in the State administration spoke of the Governor's strong commitment to tackling climate change. A number of key policy initiatives have been developed and implemented. In 2006, the then Democratic governor, Rod Blagojevich, signed an executive order to set up the Illinois climate change advisory group. The advisory group published a report which called for a target of a 60 per cent reduction in greenhouse gases below 1990 levels by 2050.

The policy has not been without its problems. One policy advisor I spoke to talked of the concern among some groups, including the labour unions, about the potential negative economic and employment impact on sectors such as the coal industry. In January 2015, Quinn was succeeded as governor by the Republican Bruce Rauner. Rauner in his inaugural speech as governor omitted any reference to the issue of tackling climate change. Despite this however, Illinois is taking action on climate change as part of its commitment to the CPP (see Chapter 2). Under the CPP, each state is assigned its own goals and Illinois is committed to reducing greenhouse gas emissions by 34 per cent below 2012 by 2030. For the Union of Concerned Scientists (UCS), Illinois is in a good position to meet these targets given its current move away from coal generation and increased investment in renewable energy and energy efficiency (UCS 2016). As part of its commitment to tackle climate change through the renewable portfolio standard (RPS), Illinois produced 10,000,000 MWH of wind energy in 2013. The US EPA estimates that Illinois will reach 18,000,000 by 2030. Together with solar power, it is estimated that Illinois will be producing some 32,000,000 MWH by 2030 (Sierra Club 2017).

Texas

The state of Texas makes an interesting case study. Whilst it has made investments in renewable energy, in particular wind power, it has actively opposed efforts by the federal EPA to regulate greenhouse gas emissions. Indeed since 2009 Texas has taken legal action on 22 occasions to oppose new regulations on controlling greenhouse emissions. However, Arevalo argues that 'public beliefs on environmental policy are not reflected ... in the state legislature's actions' (Arevalo 2015). Indeed, a poll conducted by Yale University found that only 14 per cent of Texans rejected the idea of global warming. More than half of those surveyed felt action should be taken at federal and state level to tackle global warming. However, only 44 per cent of those surveyed believed it was man made (Yale University 2013).

In October 2015, Texas was part of a coalition of states that resorted to legal action to block the CPP (see Chapter 2). For the state's attorney general, Ken Paxton, the federal government through the CPP 'has yet again proven its readiness to sacrifice American jobs' in order to promote its 'liberal agenda' (Arevalo 2015). For Luke Metzger, director of environment Texas, a citizens-based environmental advocacy group, the approach of Texas is just one more example of a political failure to hold big business to account. 'The core problem' argues Metzger is 'that powerful industries spend millions of dollars on campaigning and lobbying and the result is that politicians are catering to the interests of big donors' (Arevalo 2015). Such a viewpoint reflects the elite model of power that I discussed in Chapter 3. Indeed industrial giants Calpine and Royal Dutch Shell were among a number of corporations which backed the litigation against the CPP.

But this support of big business simply ignores the reality of climate change and the need to take action. A recent study by the risky business project points to the likelihood of a sharp increase in heat related deaths and coastal storm damage if no action is taken to mitigate climate change. Founders of the risky business project include former US treasury secretary Henry Paulson, former New York City mayor Michael Bloomberg and billionaire hedge fund manager turned environmental campaigner Tom Steyer. They argue that Texas will be one of the areas worse affected by climate change unless substantive action is taken (Risky Business Project 2015).

Pennsylvania

In the rustbelt state of Pennsylvania, action on climate change has developed over the last decade. A 2008 law, the Pennsylvania climate change act directed the state Department of Environmental Protection (DEP) to conduct an overall climate change impact assessment and to develop an action plan to reduce green-house gas emissions. This led to the publication of a CAP in December 2009. In the forward to the CAP, the DEP argued that 'Anthropogenic climate change and the increase in greenhouse gas emissions is a very real challenge facing each of us'. The CAP set a target of a 30 per cent reduction in greenhouse gas emissions from 2000 levels by 2020 (Pennsylvania DEP 2009). The DEP set out an updated CAP in 2015. It sets out strategy to tackle climate change involving 13 work plans across a number of sectors including building design, renewable energy and energy efficiency. It sets out energy consumption reductions of 80 per cent in new buildings and 50 per cent in existing buildings by 2030. The CAP analyses state greenhouse gas emissions from the base year of 2000 through to 2012. It shows an 11 per cent reduction in emissions (Pennsylvania DEP 2016). The current governor of Pennsylvania is the Democrat, Tom Wolf. He has been an outspoken critic of the actions of President Trump to roll back the climate change policies of the Obama administration.

American cities and the fight against climate change

Cohen and Millar make reference to the 'far sighted climate change policies of a number of US cities' (Cohen and Miller 2012: 47). In similar vein, Selvin and Van Deer talk of a variety of US cities which have put in place strategies to tackle climate change and to reduce greenhouse gas emissions (Selvin and Van Deer 2009: 5). Furthermore as Krause notes 'a considerable – and some say surprisingly large – amount of action to reduce greenhouse gas emissions is being taken by subnational governments' (including cities) (Krause 2015: 82). The US EPA has in place a local climate and energy programme to support cities and localities (EPA 2012a). As the EPA notes, US mayors and cities together with local organisations are taking a lead role in shaping policies and putting in place programmes to tackle climate change (EPA 2012b). City governments in the US have direct policy responsibility, independent of the federal government in Washington, over areas such as public transportation, energy efficiency and renewable energy. Thus cities have the potential to develop tailored policy initiatives to deal with local problems and challenges. However, it is important not to overstate the case since there are a number of cities who have 'lagged behind' in efforts to tackle climate change (Brewer 2014: 117).

There have been a number of policy initiatives at the city level which have attempted to tackle climate change and shape the broader sustainability agenda. These include the introduction of local climate action plans. The cities involved range from New York on the eastern seaboard and Chicago in the Midwest through to San Francisco on the west coast. The city of San Francisco came first in the US and Canada green city index with New York, Seattle, Denver and Boston rounding off the top five US cities. The index analyses the environmental sustainability of 27 major cities in both countries in areas such as carbon dioxide emissions, air quality, transport, energy and environmental governance (Economist Intelligence Unit 2011). A broader green index in 2012 looked at the performance of cities across five continents. The results suggested that US cities had still some way to go however. For example, they had higher per capita greenhouse gas emissions than Europe and Asia combined. US cities had higher water consumption rates but scored well on recycling (Economist Intelligence Unit 2012). The sustainable cities index (SCI) ranks the top 100 cities in terms of a number of criteria. These include energy use, greenhouse gas emissions, air pollution, education, environmental risk, green spaces, education and health. Top of the list was the Swiss city of Zurich. The highest placed US city was New York which was ranked 26. The other US cites to make the top 50 were Boston (34), San Francisco (38) and Seattle (43) (SCI 2016).

A number of US cities played a key role in what was known as the Chicago climate exchange. Set up in 2003, it brought together private sector companies, cities and municipalities, universities and utilities. It was a voluntary but legally binding greenhouse gas reduction and trading system. Based on the principles of the Kyoto Protocol it set targets for greenhouse gas reductions amongst its constituent parts. At its height it had in the region of 400 members. However, it came to an abrupt halt in 2010. A number of factors played a role in this, principle of which was a big fall in the price of carbon (key to its trading system) and the failure of the US Congress to pass a binding cap and trade bill. A cap and trade bill at the federal level would have set the policy context for schemes such as the Chicago climate exchange. The failure of Congress to act in this area represented a missed opportunity.

Other initiatives have been more long lasting. The US conference of mayors has been active in shaping policies to tackle climate change at the local level. In 2005 city mayors from across the US signed the *climate protection agreement* (CPA). The CPA was based on the UN Kyoto Protocol on greenhouse gas reduction. President George W. Bush whilst in office had actively opposed the Kyoto

Protocol. But as Farley and Smith argue, 'In the absence of national leadership to ratify the Kyoto Protocol, cities' leaders (who signed up to the CPA) have committed themselves to enacting local programmes to reduce greenhouse gas emissions as well lobbying politicians at the federal level to support local climate change agendas (Farley and Smith 2015: 99). Under the Agreement, participating cities committed themselves to strive to meet or beat the Kyoto Protocol targets in their own communities and to urge state and federal governments to do likewise.

To date some 1060 city mayors have signed up to the CPA (US Mayors 2017). Research by Lutsey and Sperling (2008) estimate that if all US cities had signed up to the CPA, the outcome would have been a reduction of 9 per cent in total US emissions, a relatively small sum set against a target of a global reduction of greenhouse gas emissions of 80 per cent by 2050. However, as Krause argues, action by city governments to tackle climate change can act as a catalyst for broader initiatives to tackle climate change (Krause 2015: 100), cities can lead by example. A decision to install solar panels of the roof of City Hall might encourage local businesses to do likewise. In a further initiative in 2016, mayors from across the US came to form the mayors' climate action agenda. It commits mayors to work together to strengthen local efforts for reducing greenhouse gas emissions and to supporting efforts for binding federal and global level policy making.

A recent study analysed the greenhouse gas emissions data of 116 cities across the US, representing some 14 per cent of the country's population. It found that 37 cities have set a target to reduce emissions by 80 per cent by 2050. 62 cities set a target between 2020 and 2030, equal to or greater in ambition than the US national target of a 26 to 28 per cent reduction by 2025. The study talks of the important impact of such policies and the leadership shown by many cities but argues that they are only scratching the surface. It concludes that there are still many cities yet to see the benefits of action on climate change (ICLEI 2015).

Cohen and Millar have argued that 'cities tend to be free from the heightened political polarisation seen at the Federal level' in respect of policies to tackle climate change and to promote the broader sustainability agenda (Cohen and Miller 2012: 45). Indeed there is considerable validity in such a view. Increased partisanship in Washington, DC has led to policy gridlock in key areas of policy making, including attempts to put legislation on the statute book to tackle climate change. There is a much less febrile political atmosphere at the local city level. However, we need a word of caution here. There still exist significant political disagreements between and within political parties at the local level on policy areas, the result of a complex interplay of vested interests, policy saliency and other factors.

US cities and climate change policy: some case studies

A large number of US cities have made public commitments to tackle climate change. Chicago, for example, has set out a strategy to reduce its greenhouse gas emissions by 80 per cent below 1990 levels by 2050. Other cities taking the lead on climate change include Los Angeles and New York. However, it is important to note that there is considerable variety in the approaches taken by US cities to tackle climate change and reduce greenhouse gases. Some cities focus specifically on attempts to reduce greenhouse gas emissions from their own municipal operations whilst others have adopted a broader community focus. Research by Bae and Feiock (2013) in a survey of 679 US cities found that the focus on internal operations is significantly more numerous than those with a broad community focus. Greenhouse gas emissions inventories and CAPs are two key elements of the tool kit that cities employ. A greenhouse gas emissions inventory involves the identification and measuring of emissions from a designated entity or physical area. CAPs set out broad strategy and detail specific emissions reduction targets and the means by which these targets will be met. In any successful city initiative to tackle climate change, political leadership and community participation and buy in are essential. It is also important to frame action on climate change in a way that stresses the co benefits that can come about, such as better air quality and the creation of green jobs. This is an important element for local politicians as they seek to obtain voter buy in. I will now focus on the climate change initiatives of some specific US cities.

New York

New York has taken a particular policy stance. In its climate action programme it states that 'Climate change must be considered in all short term and long term infrastructure and policy planning initiatives' (New York City 2008). Its CAP is part of its larger PlaNYC initiative which was launched in 2007 by the then Mayor of New York, Michael Bloomberg. Its stated aim is to adopt a city wide approach which seeks to strengthen the economy, prepare for climate change and improve the quality of life for New York. Per capita greenhouse gas emissions in New York are significantly lower than the US average and have a distinct composition. The operation of buildings creates a far larger percentage (around 75 per cent) and transportation a far lower percentage (around 21 per cent) of emissions than in most cities and the US as whole. In 2009, the city administration introduced 'ambitious building energy efficiency legislation' covering 22,000 buildings which accounted for 45 per cent of New York's greenhouse gas emissions (Cohen and Miller 2012: 45). By 2012, net emissions in the City were down by 16.1 per cent, this despite an increase in the number of buildings and people. In particular, the use of less carbon intensive methods of electricity generation accounted for just over 11 per cent of net emission reduction. As Krause notes, the impact of PlaNYC appears to be quite significant (Krause 2015). Looking to the future New York City has set a target of a 30 per cent reduction in greenhouse gases by 2030 (with 2005 as the baseline). In its 2008 CAP, it states that 'Climate change must be considered in all short term and long term infrastructure and policy planning initiatives' (New York City 2008). In 2014, it went a step further with a target of an 80 per cent reduction in greenhouse gases by 2050, with more energy efficient and sustainable buildings a key element of the strategy. For New York city mayor, Bill de Blasio, 'Global climate change is the challenge of our generation' (New York City 2014).

Chicago, Illinois

The city of Chicago adopted its CAP in 2008 (Chicago 2008) The CAP set out a two stage strategy for reducing emissions, a medium term goal of reducing emissions to 25 per cent below 1990 levels by 2020, together with a longer term goal of an 80 per cent reduction by 2050.

The Chicago CAP is based around five key strategies: increasing energy efficiency in buildings, increasing the production and use of renewable energy, improving transportation, reducing waste and industrial pollution, and preparing for climate adaption (Chicago 2012). Political leadership plays an important role in the strategy adopted by Chicago. The former Mayor, Richard Daly, provided the initial impetus to get the CAP off the ground. His stated aim was to make Chicago the most environmentally friendly city in the nation. As Krause notes 'The strong influence of a powerful political entrepreneur (Mayor Richard Daly) who acted as a champion for the plan made it easier to bring together city departments and community leaders and secure necessary resources for the project' (Krause 2015: 98). The city's climate policy has also benefited from strong community support from a variety of local interests including labour unions, the business sector, the city's universities and the non-profit sector. Indeed, community groups have a formal role in the planning process, serving on a task force that advises the mayor and the city of Chicago. The current mayor of Chicago, Rahm Emanuel, has stated that he wants 'Chicago to be the greenest city in the world' and goes on to say that he is 'committed to foster opportunities for Chicagoans to make sustainability a part of their lives and their experiences in the city' (Chicago 2015b).

Co benefits were a key driver behind the effective development and roll out of the Chicago CAP. The CAP laid a strong emphasis on the co benefits of tackling climate change, such as cost savings (through increased energy efficiency), improved air quality and the boost to the local economy through such mechanisms as the creation of green jobs. Such a framing of action on climate change is an important element in securing community support and 'may help justify the plan to members of the public who may be sceptical' (Krause 2015: 99). In 2015, Chicago published its sustainable Chicago action plan. Its stated aim was to 'Establish Chicago as hub for the growing sustainable economy'. The plan restated the CAP targets on greenhouse gas reduction in the city with a number of detailed policy proposals. These included: improving city wide energy efficiency by 5 per cent, improving overall energy efficiency in municipal buildings by 10 per cent and reducing municipal fossil use by 10 per cent (Chicago 2015a).

Los Angeles

The city of Los Angeles, the epitome of the car culture, faces many environmental challenges. As Mazmanian notes, 'From the very start, Los Angeles has had the unenviable distinction of being the most heavily polluted community in the nation' (Mazmanian 2009). However, the city has put in place a number of policies in an attempt to tackle the issue of air quality and to take action on climate change. For the city administration 'Los Angeles is taking innovative approaches to dealing with the issue of climate change' in order 'to create a more environmentally friendly, conscious way of life' (Los Angeles 2017). The plans are among the most ambitious in the US. It has put in place a CAP. It also has a broader sustainability strategy. The mayor of the city has a sustainability office and there is also a chief sustainability officer. Los Angeles also has more solar power than any other city in the US. It has also embarked on a major programme of investment in public transport.

In 2015, Los Angeles launched its sustainable city plan. Introducing the plan, the Democratic mayor Eric Garcetti argued that it was 'not just an environmental issue – by addressing the environment, economy and equity together, we will move forward toward a truly sustainable future'. He goes on to argue that 'Sustainability is not a narrowcast endeavour' but will 'permeate everything we do in the city' (Los Angeles 2015). For the mayor, the Los Angeles strategy sets a course for a cleaner environment and a stronger economy, with a commitment to equity at its foundation.

Using 1990 as the baseline, the plan sets targets to reduce greenhouse gas emissions by 45 per cent by 2025, 60 per cent by 2035 and 80 per cent by 2050. To meet these targets the plan has set in place a number of policies across different sectors. With more than 250 days of sunshine per year, Los Angeles has set out a strategy to increase the total megawatts (MW) of local solar photovoltaic (PV) power. It has set a target of between 900 and 1500 MW of solar power by 2025 rising to between 1500 and 1800 MW by 2035. A second key focus is on energy efficiency. Buildings are the largest users of electricity in the city. The aim is to significantly reduce energy consumption per square foot across all building types in the city. Using 2013 as the baseline, a target has been set for energy reduction per square foot for all buildings of at least 14 per cent by 2025 and 30 per cent by 2035.

In addition, by 2025 the city will have completely divested from coal fire power plants, a key contributor to greenhouse gas emissions. Los Angeles also sees the creation of green jobs as a means not only to tackle climate change but to create high skilled and long term jobs in areas such as energy efficiency, public transit growth and clean energy technology. There is also a key focus on environmental justice. The sustainability plan notes that it is often low income families and communities that bear the brunt of environmental pollution and poor air quality. To start to tackle this problem, policies are being developed to improve public transport and general infrastructure planning. In addition, the Los Angeles city council has adopted the *clean up, green up* ordinance which targets the most polluted neighbourhoods through the piloting of green zones (Los Angeles, 2015).

Portland, Oregon

The Pacific North West coast city of Portland is a good example of an early adopter. In 1992 it became the first city in the US to set out a CAP. Since 1990, total carbon emissions in the city have declined by 14 per cent, whilst at the same time some 75,000 jobs have been added to the local economy, together with a 31 per cent growth in population. The city set out its most recent CAP in 2015. It stresses the strategic importance of the widest participation by all stake-holders in the battle to tackle climate change. The CAP has set a target of a reduction in greenhouses gas of 80 per cent below 1990 levels by the year of 2050, together with an interim target of a 40 per cent reduction by 2030. Action on climate change in Portland is part of a broader sustainability agenda with a focus on the creation of high skill green jobs, well being, improved health, equity, inclusiveness and social justice (Portland 2015).

Austin, Texas

Austin is a Democrat leaning and liberal city (with a Democratic mayor Steve Adler) within the conservative and Republican state of Texas. Texas, historically the oil centre of the US, has seen something of an oil boom over the last two years brought about by the hydraulic fracking revolution. The city of Austin, the state capital of Texas, set out its Community Climate Plan (CPPL) in June 2015. It talks about making the city a global leader in meeting the challenges faced by climate change. The document stresses the key role of city leadership in tackling climate change, together with the need to work in partnership with the private and non-profit sectors. The CPP also recognises the crucial importance of the widest possible community involvement in Austin. The CPP aims to achieve net zero community wide greenhouse gas emissions by 2050 and sets out a blueprint for action. There are also a set of intermediate targets for 2020, 2030 and 2040. The plan is part of a broader sustainability strategy which stresses the import-ance of the stewardship of the natural environment. There is also a strong emphasis on the importance of community well being (Austin 2015) as well as a dedicated office of sustainability.

Kansas City, Missouri

Although as Krause notes 'it is unlikely to be described as a pioneer' in tack-ling climate change at the city level, Kansas City has nonetheless taken signi-ficant steps in an attempt to reduce greenhouse gas emissions. The first major initiative can be traced back to 2005 when the then mayor, Democrat Kay Barnes, signed the Mayor's climate protection agreement. This resulted in the setting up of a steering community made up of community leaders, one of whose principal tasks was to make recommendations for actions to be included in the city's CAP. Key members of the steering group included representatives from the local utility company, a number of not-for-profit organisations and the metropolitan planning organisation and the local chamber of commerce. As Krause notes, such groups were chosen as they 'represented stakeholder groups whose support would be necessary' if there was to be a successful outcome to the climate action policy (Krause 2015: 93). This links to the earlier point I made about getting community buy in if local action on climate change is to be successful. In 2008, the CAP was put into place. It contained two key elements. First, there was a focus was on reducing greenhouse gas emissions emanating from the city council's own internal operations. The CAP set out a targeted reduction of 30 per cent below the levels of those in 2000 by 2020. Second, similar targets for reducing greenhouse gases were set for the wider community covered by the area of the Kansas City Council (Kansas City 2008). Since 2011, the mayor of Kansas City has been Sly James, who is not aligned to any political party. He was one 38 local mayors who wrote an open letter to President elect Donald Trump on 22 November 2016 stressing the importance of action on climate change.

US cities and climate action: some concluding thoughts

In the above section I have looked at a number of local climate change policies. As we have seen, around 1060 city mayors have signed up to local action on climate change. But this still leaves a large number of cities with little or no action on climate change (Foss and Howard 2016). There are a number of possible explanations for this. It might be fair to say that individual city action on climate change might be seen to run counter to assumptions about collective action problems (Olson 1971). Why in the face of a policy challenge, with all the opportunities for free riding, do some cities take action and commit resources to reduce greenhouse gases?

Ideological factors offer one explanation for cities that commit themselves to tackling climate change and developing a broader sustainability agenda. Foss and Howard argue that 'cities leading on climate change action tend to be located in politically liberal (and Democratic) regions and states' and that this 'reflects political polarisation on climate change' (Foss and Howard 2016). A study by Foss and Howard of the 100 largest cities in the Dallas Fort Worth region of Texas, a politically conservative metropolitan area, found that only 18 per cent had any type of actions or programmes to tackle climate change. However, the same study did find that 66 per cent of these cities had introduced in-house energy efficiency initiatives. 47 per cent of the cities surveyed had developed policies to encourage domestic residents and local business to be more energy efficient (Foss and Howard 2016).

Interest group theory also offers us an insight. Some surveys have found that the presence of a large number of environmental lobby groups in a city is likely to increase the possibility of city government action on climate change. On a more directed policy level, we have the notion of the co benefits that can and do result from action on climate change. These can include the reduction of local air pollution and less traffic congestion as result of improved local public transport systems. As Brewer notes, 'some local leaders understand more clearly and/or value more highly the co benefits of climate action' (Brewer 2014: 123). For example, in the case of the city of Chicago, co benefits are a key element in its action on climate change. As Krause notes, 'They receive considerable emphasis in publicly released communications'. Stated co benefits include improved air water quality, economic development and more attractive communities to live in (Krause 2015: 99). The climate action plan of the city of Austin has an emphasis on co benefits, such as improving community well being (Austin 2015). The city of Portland climate action plan places an emphasis on co benefits such as the creation of high quality green jobs and improved health outcomes (Portland 2015). For Foss and Howard, energy efficiency measures 'can easily be framed in terms of economic savings, making them more politically viable in conservative areas' (Foss and Howard 2016).

Conclusion

This chapter has argued that relative lack of action at the federal level of US government to tackle the global threat presented by climate change (in particular at the legislative level) has opened up a policy space which a number of US states and cities have sought to fill. We have noted how a number of states and cities have put in place substantive policies to tackle climate change. Some have gone further and developed action on climate change in the context of a broader sustainability agenda. However, a word of caution is needed here as the overall picture is somewhat patchy with many states and cities taking limited or no action on climate change. There is an ideological divide here. Democrat leaning areas tend to be more active on climate change than their Republican counterparts. This is reflective of a long standing political polarisation on the issue of climate change which can be seen in politics in Washington, DC and in public opinion. Yet taken together, climate policy action at the state and city levels can and does have a meaningful impact.

References

Arevalo, A. (2015) 'Texas Climate Policy Must Reflect Popular Culture' *Daily Texan*, 5 November.

Austin (2015) *Austin Community Climate Plan* City of Austin, TX.

Bae, J. and Feiock, R. (2013) 'Forms of Government and Climate Change Policies in US Cities' *Urban Studies*, vol. 50, issue 4, pp. 776–788.

Brewer, T. (2014) *The United States in a Warming World*, Cambridge: Cambridge University Press.

Breggin, L. (2012) 'Building Building Energy Codes' *Environmental Forum*, vol. 29, issue 4, pp. 7–13.

Chicago (2008) *Climate Action Plan* City of Chicago.

Chicago (2012) *Sustainable Chicago Action Agenda, 2012–2015, Highlights and Look Ahead*, City of Chicago.

Cohen, S. and Miller, A. (2012) 'Climate Change 2011: A Status Report on US Policy' *Bulletin of the Atomic Sciences*, vol. 68, issue 1, pp. 39–49.

Economist, The (2016) 'Washington State's Carbon Tax of Wood and Trees'.

Economist Intelligence Unit (2011) 'Us and Canada Green City Index' Munich: Siemens GmBH.

Economist Intelligence Unit (2012) 'The Green City Index' Munich: Siemens GmBH.

Farley, F. and Smith, Z. (2015) *Sustainability: If it's Everything is it Nothing?*, London: Routledge.

ICLEI (2015) *Measuring up 2015: How US Cities are Accelerating Towards National Climate Goals* International Council for Local Environmental Initiatives, www.iclei.org.

Kansas City (2008) *Climate Action Plan*, Kansas City, MO.

Krause, R. (2015) 'Climate Policy Innovation in American Cities' in Wolinsky-Nahmias, Y. (ed.), *Changing Climate Politics: US Policies and Civic Action*, Los Angeles: Sage.

Los Angeles (2015) *Transforming Los Angeles: Environment, Economy, Equity*, City of Los Angeles, CA.

Lutsey, N. and Sperling, D. (2008) 'America's Bottom up Climate Mitigation Policy' *Energy Policy*, vol. 36, issue 2, pp. 673–685.

Mazmanian, D. (2009) 'Los Angeles' Clean Air Saga: Spanning the Three Epochs' in Mazmanian, D. and Kraft, M. (eds), *Towards Sustainable Communities: Transition and Transformations in Environmental Policy*, Cambridge, MA: MIT Press.

New York City (2008) *Climate Change Program Assessment and Action Plan* Department of Environmental Protection Action Plan, New York City.

New York City (2014) *One City Built to Last: Transforming New York City Buildings for a Low Carbon Future* New York City.

Olson, M. (1971) *The Logic of Collective Action: Public Goods and the Theory of Group*, Cambridge, MA: Harvard University Press.

Pennsylvania DEP (2009) *Climate Change Action Plan* State of Pennsylvania.

Pennsylvania DEP (2016) *2015 Climate Change Action Plan: Updated August 2016* State of Pennsylvania.

Portland (2015) *Climate Action Plan* City of Portland, WA.

Rabe, B. (2015) 'A New Era in States' Climate Policies' in Wolinssky-Nahmias, Y. (ed.), *Climate Change Politics: US Policies and Civic Action*, Los Angeles: Sage.

Selvin, H. and Van Deer, S. (eds). (2009) *Changing Climates in North American Politics: Institutions, Policy Making and Multi-level Governance*, Cambridge, MA: MIT Press.

Sussman, G. and Daynes, B. (2013) *US Politics and Climate Change: Science Confronts Policy*, Boulder, CO: Lynne Rienner Publishers.

Yale University (2013) 'Climate Change in the Texan Mind' Yale programme on Climate Change communication, 23 September.

Internet sources

ACEEE (2016) 'Research Report'. Available at: 2016 http://aceee.org/state-energy-efficiency-scorecard, accessed 19/04/17.

C2ES (2016) 'Climate Action Plans'. Available at: www.c2es.org/us-states-regions/policy-maps/climate-action-plans, accessed 19/03/17.

Chicago (2015) 'Chicago Climate Action Plan'. Available at: www.chicagoclimateaction.org, accessed 10/06/15.

Chicago (2015) 'Environment and Sustainability'. Available at: www.cityofchicago.org/city/en/progs/eenv/html, accessed 04/04/17.

EPA (2012a) 'Local Climate and Energy Program'. Available at: www.epa.gov/state-localclimate/local, accessed 18/01/12.

EPA (2012b) 'Local Governments'. Available at: *www.epa.gov/region9/climatechange/localgovernment,* accessed 18/01/12.

Foss, W. and Howard, J. (2016) 'Conservative Texas Cities Finds Many Lagging or Faltering'. Available at: http://blogs.lse.ac.uk/usappblog/2016/03/16/although-cities-often-are-touted-as-climate-change-policy-leaders-a-close-look-at-politically-conservative-texas-cities-finds-many-lagging-or-faltering/, accessed 24/03/17.

Los Angeles (2017) 'Los Angeles Regional Collaborative for Climate Action and Sustainability'. Available at: www.lasregionalcollaberative.com/policy-plans-climate-change, accessed 20/03/17.

SCI (2016) 'Sustainable Cities Index'. Available at: www.arcadis.com/en/united-states/our-perspectives/2016/2016-global-sustainable-cities-indexputting-people-at-the-heart-of-city-sustainability/, accessed 04/08/17.

Sierra Club (2017) Available at: https://content.sierraclub.org/creative-archive/sites/content.sierraclub.org.creative-archive/files/pdfs/0796%20Illinois%20CPP%20Fact%20Sheet_01_web_2.pdf, accessed 04/04/17.

Risky Business Project (2015) Available at: https://riskybusiness.org/, accessed 04/03/17.

UCS (2016) 'Meeting the CPP in Illinois'. Available at: www.ucsusa.org/clean-energy/increase-renewable-energy/clean-power-plan-illinois#.WRsO0oWcHIU, accessed 04/03/17.

US Mayors (2017) 'The US Conference of Mayors Climate Protection Agreement'. Available at: www.usmayors.org/mayors-climate-protection-center, accessed 04/05/17.

6 The US as a world leader in tackling climate change and building a more sustainable world

Opportunities and constraints

Introduction

This chapter will argue that although the US has historically played a central role in creating the current challenge that the planet faces with climate change, it also has the potential to be a world leader in developing strategies to tackle climate change. Whilst previous chapters have highlighted the tensions with regard to climate change policy, there have nonetheless been a number of substantive policy developments. The chapter will argue that there is a potential to build on such developments. Yet to discuss the role of the US as world leader in tackling climate change might be seen at first glance as unlikely, if not somewhat laughable. For as Farley and Smith note, the US has been frequently criticised for its lack of effort in leading and supporting global efforts tackle issues such as climate change (Farley and Smith 2014: 94). Indeed, right wing conservative think tanks, backed by big oil, have pushed a climate change denial agenda which has considerable traction amongst significant sections of the voting public. Mainstream media coverage of the causes and consequences of climate change has been patchy to say the least. Under the Presidency of George W. Bush from 2001, the US failed to engage with global agreements to tackle climate change. The failure of the Bush administration to ratify the 1997 Kyoto Protocol is the most striking example of this.

The election to the Presidency of Barack Obama in November 2008 did, however, seem to point to a new era in tackling climate change at the federal level in the US. The White House, as we noted in Chapter 2, launched a series of climate policy initiatives during the eight years of the Obama administration. However, Obama had to operate in a toxic political climate of partisan politics whereby the Republican Party in Congress sought to undermine his climate change policy agenda at every turn. As a consequence no substantive policy on tackling climate change has been passed by the Congress since 2011. During his time in office Obama became increasingly reliant on executive orders and executive actions to push through his policy agenda. The Conservative leaning Supreme Court also proved to be a thorn in his side, blocking or delaying a number of the President's climate change policy initiatives including the CPP.

Yet despite all of this, this chapter will argue that the US has the potential and indeed the obligation to play a major role as world leader in tackling climate change and shaping a more sustainable world. In short, it is a big part of the problem but also needs to be a big part of the solution. For, as Farley and Smith observe, the US has both significant technical and scientific capability as well as considerable political influence (Farley and Smith 2014: 95).

The obligation of the US to tackle climate change and shape a more sustainable world

Dernbach gets to the nub of the issue arguing that the global ecological and economic footprint of the US 'is so large that it is difficult to imagine how the world can achieve sustainability unless the United States does' (Dernbach 2009: 5). Building on this theme, Jones argues that 'Americans have been the world's most profligate energy consumers for more than a decade' (Jones 2014: 1). Indeed, the ecological footprint of US on its own far exceeds the regenerative capacity of the planet.

Data from the web site Scientific American shows that the US is responsible for 25 per cent of all global consumption. The US with 5 per cent of the world's population consumes 30 per cent of its paper, 25 per cent of its oil, 27 per cent of its aluminium and 19 per cent of its copper. The average American will drain as many resources as 35 citizens of India and consumes 53 times more goods and services than someone in China. A good example of the country's ecological footprint is the fact that the average US child will create 13 times more ecological damage as a child in Brazil. More broadly, if everyone in the world lived the same lifestyle as the average American, we would need five planet earths (Scientific American 2017).

As we noted in Chapter 1, the US is the second largest emitter of global greenhouse gases. Latest figures for 2014 show that the US is responsible for 15 per cent of total global greenhouse gases (see Table 6.1). Whilst it is true that China tops the list with 30 per cent, taken on a per capita basis the US produces twice as many emissions as compared to China. Overall four countries are responsible for 61 per cent of all emissions. On a more positive note, figures from the EPA show that in the year 2014/2015, US greenhouse gas emissions fell by 2.2 per cent, in large part a result of less consumption of coal and more use of shale gas (EPA 2017). Yet it is clear that no global strategy to address climate change can succeed without substantial reductions in the emissions of the US.

In 2014, a joint report was published by the Climate Action Network Europe and Germanwatch, a public policy think tank with offices in Bonn and Berlin.

Table 6.1 Percentage of global greenhouse gas emissions by country 2014

China	30%
US	15%
European Union	10%
India	6.5%

Source: PBL, Netherlands, 2015.

The report included a climate change performance index (CCPI) which analysed how individual countries are contributing to greenhouse gas emissions and climate change, and what measures they are taking to deal with these challenges. The index is comprised of the 58 countries which emit the largest quantity of greenhouse gases. Taken together, these 58 countries account for 90 per cent of total global carbon emissions. Within the index each country is ranked based on its carbon emissions trends, its energy efficiency, its policies on renewable energy and its overall climate strategy. The US was ranked 43rd out of 58 (Climate Action Network 2014).

Another key area which impacts on tackling climate change and shaping a more sustainable world is the issue of military expenditure and associated war and conflict. Here the US stands way ahead of the field. Research by the Stockholm International Peace Research Institute (SPIRI) shows that in 2015, the US accounted for 36 per cent of all global military expenditure – see Table 6.2.

In February 2017, the newly elected President Donald Trump announced plans for a $54 billion rise in US military spending, this despite the fact that the US spends as much in this area as the next four largest spenders combined. The US military uses more oil than any other institution in the world. In total it consumes some 100 million barrels of oil a year to power ships, vehicles, and aircraft and to support ground operations (Union of Concerned Scientists 2017). Militarism and warfare pose a major threat to both people and planet.

Principle 24 of the 1992 UN Conference on Environment and Development (UNCED), the so-called Rio Summit, states that 'warfare is inherently destructive of sustainable development'. It goes on to argue that 'States shall therefore respect international law providing protection for the environment in times of armed conflict' (UNCED 1992). There are a plethora of climate change and sustainability impacts as a consequence of war and conflict. These include the destruction of forests and woodlands (which act as carbon sinks absorbing greenhouse gases), large increases in carbon dioxide emissions and water supply contamination. The US devours vast amounts of fuel in its military endeavours. As one illustration, the US military consumed in the region of one 1.2 million barrels of fuel each month during 2008 alone (Farley and Smith 2014: 97). Thus as Farley and Smith observe 'the impacts of conflict on foreign soil are not only humanitarian but also environmental and hinder sustainable development (including the fight against climate change) in developing nations' such as Iraq and Afghanistan (Farley and Smith 2014: 97).

Table 6.2 Global military expenditure by country 2015 as a percentage

US	36%
China	13%
Russia	4%
UK	3.3%

Source: SPIRI, 2015.

It is clear therefore that the US can play a pivotal role as a global leader in combating climate change and building a more sustainable world both by avoiding war and conflict itself and by using its good offices, utilising both hard and soft power to promote peace and reduce conflict.

The US as a world leader in tackling climate change: opportunities and challenges

Farley and Smith have argued that 'without participation, and leadership from the United States, it is unlikely that the global agenda for sustainable development (including strategies to tackle climate change) will be successful' (Farley and Smith 2014: 95). Furthermore, Dernbach cryptically argues that 'we (the US) can lead or follow but we are too big to get out of the way' (Dernbach 2009: 5). The US has in place attributes needed to play a leading role tackling climate change on the global stage. It is renowned for its innovation, dynamism and high level of technical skills. Silicon Valley is the exemplar of this. The US has the ability to attract the finest talent from all over the world. Whilst it is by no means axiomatic that such ingredients are of themselves a guarantee of effective American global leadership on climate change, they do nonetheless provide a very important and necessary starting point. When mixed together with strong leadership, we have a potential recipe for success. A good example of this is the *Breakthrough Energy Coalition*, recently set up by the tech entrepreneurs Jeff Bezos, Richard Branson, Bill Gates and Mark Zuckerburg. For Zuckerburg 'Solving the clean energy problem is an essential part of a better world' and 'a stable climate' (Lapowsky 2017). The stated strategy is to fund start ups in a range of industries from agriculture to transport to electricity storage. The entrepreneurs will focus their investments in countries that are part of *Mission Innovation*, a consortium of 20 countries, committed to doubling investment in clear energy over five years. Such a technological approach can make an important contribution to action on climate change but it is only part of the story.

In the cold war with the Soviet Union the US used both hard power and soft power in its ideological battle with the Soviet Union. Nearly 30 years on from the end of the cold war in what one might call the 'climate war', both hard power and soft power offer the US the potential to play a positive role in shaping global efforts to tackle climate change. Aspects of hard power include economic leverage, military might, trade and political leadership and diplomacy. Soft power includes cultural influences such as the arts, television and films (for example Al Gore's film 'Inconvenient Truth'), scientific and technical capability, and the global reach of what one could term American English. All these aspects of power can be utilised by the US in developing a global role in the fight against climate change. What is required is political will and political imagination.

Political leadership is a key element in action on climate change. Early on in his Presidency, Barack Obama stressed the importance of US political leadership in the global fight against climate change. In September 2014, the US Secretary of State John Kerry promised ahead of a UN summit on climate change in New

York to put climate change 'front and centre' in American diplomacy. Kerry argued that climate change was 'an enormous challenge and this is why the United States is prepared to take the lead in order to bring other nations to the table'. Kerry further stated that he was 'personally committed to making sure' that the issue of climate change was 'front and centre of all our diplomatic efforts' (Goldenburg and Harvey 2014). Henry Paulson, secretary of the treasury under President George W. Bush, and a Republican has also written of the key role that the US should play in the fight against climate change. For Paulson, 'This is a crisis we cannot afford to ignore'. Whilst recognising that 'the United States can't solve this problem alone' Paulson argues that 'we're not going to be able to persuade other big carbon polluters to take the urgent action that's needed if we are not doing everything we can to slow our carbon emissions and mitigate our risks' (Paulson 2014). Taking up the theme, some senior Republicans in the Climate Leadership Council (CLC) recognise that the 'mounting evidence of climate change is growing too strong to ignore'. Key figures in the CLC include George Schultz, Secretary of State under Ronald Reagan, and James Baker, who was Secretary of State under President George H. Bush. The CLC has called for a tax on carbon emissions, starting at $40 per ton and rising over time (CLC 2017).

And yet there are some real challenges facing politicians pushing action on climate change. In her memoir, Hillary Clinton speaks of how tackling climate change is 'a hard political sell' which requires 'bold leadership and international cooperation' (Clinton 2014: 491). In this context, she makes reference to the UNFCCC gathering at Copenhagen in 2009. As we noted in Chapter 2, for a number of observers the outcome at Copenhagen fell short of expectations. Clinton concedes that the failure of the Senate in 2009 to pass a climate change bill undermined the position of the US and led to only limited progress at Copenhagen (Clinton 2014: 505).

Clinton talks of the difficult 'choices and trade offs' that faced political leaders in seeking to reach an agreement in Copenhagen. Indeed the challenges that face political leaders in the US and more globally in the battle to tackle climate change cannot be underestimated. First of all, there is the nature of the problem. As we noted in Chapter 1, climate change is an example of what has been termed 'a wicked problem' (Sedlacko and Martinuzzi 2012). Jordan *et al.* make the important point about how climate change is 'Complex, unprecedented, and its worst impacts will be felt by people we won't meet, decades into the future' (Jordan *et al.* 2011: 122). Developing policy and political responses to climate change present us with a complex and multi-faceted challenge, bringing together a plethora of issues, whether it is seeking to eradicate poverty in the developing world or achieving energy security in a country such as the US. Combating climate change requires long term political strategies. However, politicians often see matters in the short to medium term, governed by the nature of the electoral cycle. To put it bluntly, politicians in the US as elsewhere want to get elected! And yet it is the case that to achieve substantive action on climate change will need real sacrifices by the very voters politicians seek to attract to their corner. How do they persuade us as voters and citizens to consume less, to use less energy, not to fly so much and to make less use of our cars?

Such challenges are central to the role of the US in tackling climate change. Hillary Clinton talks about how 'Once in office, President Obama and I agreed that climate change represented both a significant security threat and a major test of American leadership' (Clinton 2014: 493). But she goes on to argue, somewhat sanguinely, that trying to build a 'network of global partners willing to tackle climate change' as being 'harder than herding cats' (Clinton 2014: 494). But she does not duck the key role the US has to play. She hits the nail on the head when arguing that America's ability to lead in the global arena on climate change 'hinges on what we ourselves are willing to do at home' (Clinton 2014: 505). Linked to this point Cooper has argued that in the global fight against climate change the world needs a leader and that the US needs to step up to the plate in this regard (Cooper 2015).

In this regard, in February 2012, Secretary of State, Hillary Clinton, announced a new policy programme, the Climate and Clean Air Coalition. It is government led with 51 countries involved. It also incorporates civil society and the private sector. A key focus of its work is the reduction of short lived carbon pollutants in 11 key areas including methane, black carbon and HFCs. It was established as part of an international effort to take swift action on short lived carbon pollutants, substances which have a relatively short life time in the earth's atmosphere. It supplements the work of the UNFCCC process. The US is a founding member of the Coalition.

The actions of President Trump in rolling back on policies to tackle climate change on the domestic level, and reducing America's role on the global stage (the so-called America First policy) could well serve to damage America's long term strategic interests, leaving a space at the global level for others to fill. China has been quick to recognise this. There is increasing evidence that China is seeking a global leadership role in the fight against climate change.

As Lorenzen notes 'Less than a decade ago China was routinely accused of blocking the global fight to tackle climate change' but now they are 'lining up to be the world's leading nation in tackling it' (Lorenzen 2017). As evidence of this, China's special envoy for climate change, Xie Zhenhua, in an interview with the China Daily newspaper in January 2017, argued that 'China's firm attitude to engage in global climate change action will inject confidence amid a rising division between pro globalization and anti globalization' forces. Also in January in a speech at the World Economic Forum in Davos, the Chinese President, Xi Jinping, in an indication of China's willingness to take a leading role on climate change, stressed the importance of upholding and sticking to the COP21 Paris agreement on climate change. He emphasised the economic opportunities arising from the fight against climate change in relation to the development of new technologies and the green economy. On a more strategic level he made it clear that the global sustainable development goals (SDGs) should be implemented 'to insure balanced development across the world' (Lorenzen 2017).

The Chinese President's speech was welcomed by the executive director of Greenpeace International, Jennifer Morgan. She argued that the President's 'address had not only calmed nerves but boosted global confidence' and highlighted 'China's evolving calculus towards action' on climate change. 'Now more than ever', Morgan commented 'the world needs to follow committed powers like China to safeguard and enhance the hard fought international climate regime' (Lorenzen 2017). Paradoxically, Trump's America First policy could well allow China to steal a march on the global stage, with the potential loss of a great deal of economic, political and strategic influence for the US. As Volcovici and Ling Wong observe China 'is poised to cash in on the good will it could earn by taking on leadership in dealing with what many governments is one of the most urgent issues on their agenda' (Volcovici and Wong 2015). To emphasise this point, Zou Ji, deputy director of the National Centre for Climate Change Strategy and a senior Chinese climate talks negotiator argues that 'Proactively taking action on climate change will improve China's international image and allow it to occupy the moral high ground, which will then spill over into other areas of global governance and increase China's global standing, power and leadership' (Volcovici and Wong 2015). The previous Obama administration recognised the strategic advantage to America in taking a key role in the global effort to tackle climate change and the importance of working with key international actors such as the Chinese. It is to this that I now turn.

US/China relations

The Obama administration worked closely with the Chinese government to build momentum ahead of the COP21 Paris agreement on climate change. The catalyst for this can be traced back to the summer of 2013 when Barack Obama and Chinese President Xi Jingping held an historic summit in June of that year in Palm Springs California, the aim of which was to create 'a new spirit of co-operation between the world's two economic superpowers' (Roberts and Goldenburg 2013). The two leaders agreed on a broad strategy to tackle climate change in partnership along with other countries. This was significant in itself. In the past China had argued that cutting greenhouse gas emissions to tackle climate change would compromise its economic growth whilst the position of the US was that it would not act on climate change unless China did as well. The most significant outcome of the meeting between the two presidents was an outline agreement to reduce the use of hydro fluorocarbons (HFCs) which are commonly used in things such as air conditioning systems and refrigerators. HFCs are an extremely potent source of greenhouse gas, up to 1000 times more so than carbon dioxide. The importance of increasing US/China co-operation on action on climate change was further signalled by the then Secretary of State, John Kerry, on a visit to Beijing in February 2014.

Building on the June 2013 US/China summit, a global deal to cap and reduce the use of HFCs was agreed in the Rwandan capital, Kigali, in October 2016. John Kerry, the then US Secretary of State, heralded the outcome as a 'monumental step forward' (*Guardian* 2016). For President Obama, the agreement was 'an ambitious and far reaching solution' (*Guardian* 2016). Whilst concerns were expressed by some about the pace of the deal as it is not due to start until 2019, Durwood Zaelke of the Institute of Governance and Sustainable Development took a more optimistic tone arguing that it will result in 'the largest temperature reduction ever achieved by a single agreement' (*Guardian* 2016). Whilst as we note above, HFCs are a major source of greenhouse gas, they can clear out of the atmosphere relatively quickly, on average in about 10 to 15 years. This means that cutting HFCs has the potential to deliver relatively quick results. For David Doninger, climate and clean air programme director of the NRDC, the agreement was 'equal to stopping the entire world's CO_2 emissions for more than two years' (*Guardian* 2016). In a further sign of strengthening US/China co-operation on climate, President Obama met with the Chinese President on the eve of the G20 summit in Hangzhou, China in early September, when both counties ratified the COP21 Paris climate deal.

The US and the fight against climate change: policy contradictions on the world stage

Whilst the approach to action on climate change of the Obama administration represented a step change from the Presidency of George W. Bush, there are nonetheless some apparent contradictions in the policy approach. This is particularly so in the area of international trade agreements. A case in point is the US/European Union trade deal, the Transatlantic Trade and Investment Partnership (TTIP).

TTIP was a flagship international trade policy of the Obama administration. The TTIP deal itself is currently in abeyance, this is in large part the product of concerted opposition from environmental organisations and development NGOs. Furthermore, the new elected President, Donald Trump, has made it clear that as part of his America First policy, he would not ratify any TTIP deal as it is currently constructed. Nonetheless, an analysis of the nature of TTIP does reveal some areas of concern in terms environmental protection and the fight against climate change. TTIP involves a set of free trade negotiations between the US and the European Union (EU). During his time in office, Obama made the case that globalisation, particularly in the form of free trade, had improved the lives of many millions across the globe. For the European Commission, TTIP is about driving growth and creating jobs by removing cumbersome and bureaucratic trade between the US and member states of the EU, and securing better access to markets (European Commission 2017). A key element of TTIP as originally envisaged was the investor state dispute settlement or ISDS. The ISDS may sound rather mundane, but if such a thing were to be implemented, there could major consequences for the environment and policies to tackle climate change. It would allow, for example, companies seeking to invest in the US and Europe to bypass national legal processes and to challenge national governments in special international tribunals if they felt that laws on environmental protection and climate change posed a threat to their commercial interests.

The European Commission itself argues that TTIP will establish global rules to promote and support sustainable development and to protect the environment (European Commission 2017). Obama himself, whilst acknowledging that the benefits of free trade 'are often diffuse', in a visit to Germany in April 2016 argued that 'TTIP will not lower standards, it will raise standards even higher; high standards protecting consumers to give them more choice' together with important environmental standards (Politics, 2016). Others take a different view. The development NGO, War on Want, argues that TTIP as originally envisaged would result in a race to the bottom with regard to environmental regulations and protection (War on Want 2017). Oxfam makes a similar point (Oxfam 2017). The campaigning group, Global Justice Now, argues that the deal signals a shift in power to powerful corporations and is major threat to the environment and social justice (Global Justice Now 2017).

There are a number of examples, operating under rules similar to TTIP, that would appear to back up the critics' case. In September 2013, the Canadian oil and gas company, Lone Pine Resources, initiated an ISDS against the Canadian government under the investment chapter of the North America Free Trade Area (NAFTA). The case followed a decision by the province of Quebec to introduce a moratorium on drilling under the St. Lawrence River whilst environmental and other impact studies were carried out. The company, in response, argued that the move was an 'arbitrary, capricious and illegal revocation' of its 'valuable right to mine for oil and gas' (Friends of the Earth 2016: 19). Lone Star Resources is suing the Canadian government for in excess of $110 million.

Let us look at another example. In March 2015, an arbitrations panel constituted under NAFTA ruled that a Canada environmental review process violated NAFTA's investment protection rules. Bilcon, a US company, wanted to build a large quarry and marine terminal in an ecologically sensitive area of eastern Canada. Based on the decision of an environmental assessment panel which highlighted the negative environmental impacts of the scheme, the provincial government of Nova Scotia denied approval for the scheme (Friends of the Earth 2016). The company is seeking some $300 million dollars in damages. In El Salvador, local communities successfully lobbied to block the development of a large gold mine which threatened to contaminate local water supplies. Pacific Rim, the Canadian company which applied to dig the mine is suing the government of El Salvador for $315 million. Such examples would seem to suggest a contradiction between Obama agenda on environmental protection and action on climate change and his commitment to globalised free trade.

More recently the US and the EU have sought to address the environmental concerns emanating from TTIP. In a recent joint assessment of the state of play with regard to TTIP, the EU trade commissioner, Cecilia Malmstron and the then US trade representative, Michael Froman set out plans for a new mechanism to replace the much criticised ISDS. It was to be replaced by a 'modern and transparent investment court system (ICS) that effectively protects investment while fully preserving the rights of government to regulate' (European Commission 2017). The new proposal has met with strong criticism in environmental circles.

For FOE, 'There is nothing in the proposed (new) rules that prevent companies from challenging government decisions that protect health and the environment' (Friends of the Earth 2016: 4).

The European Commission now expect TTIP to be finally agreed by 2019 or 2020 (European Commission 2017). With the election of Donald Trump as US President in November 2017, this seems very unlikely at this time. As part of his America First policy he has strongly argued that such multilateral deals are bad for the American economy and the American worker. Indeed, shortly after coming to office he announced that he was pulling out of the 12 nation Trans-Pacific Partnership (TPP). TPP, again strongly advocated by Barack Obama, is in many ways a companion deal of TTIP, and has been the subject of similar concerns among environmental groups.

Having noted some of the apparent contradictions between the Obama administration's support for trade deals on the one hand, and action on the environment and climate change on the other, I now turn to some aspects of President Obama broader sustainability strategy.

The Obama Presidency and sustainability

In a wide ranging and important speech covering issues such as social justice, climate change, poverty, gender equality and development to the UN General Assembly in September 2015, Obama spoke of the hope for the world 'that is available to us through collective action' (White House 2015a). He also made reference to the 800 million people across the world struggling to survive on less than $1–25 a day, together with the widespread risks of diseases such as malaria and HIV/AIDS.

The primary focus of Obama's address to the UN was the recently agreed global SDGs. Obama called on the global community to 'commit ourselves to the new SDGs in our goal of ending extreme poverty in the world' (White House 2015a). However, he went on to highlight the many challenges and difficulties and lay ahead in implementing the SDGs. For Obama 'many of the conflicts, the refugee crises, the military interventions over the years might have been avoided … if the wealthiest nations of Earth were better partners in working with those that are trying to lift themselves up' (White House 2015a).

Importantly, in his speech he reiterated the 'pledge' that America would remain the global leader in development. However, this is a problematic area, as many development NGOs have been critical of past policies aimed at supporting economic development in poorer countries. Such policies were based on neo liberal economic principles such as free markets, privatisation and trade liberalisation, a policy paradigm which became known as the Washington Consensus. Obama in his address did acknowledge past mistakes without being specific. But at the UN he spoke of a different approach to development now and in the future. For development to be effective, it had to be 'truly sustainable and inclusive' and based on the commitments in the SDGs (White House 2015a). Linking together action on climate change and the broader sustainability agenda, the

President spoke of how development was threatened by climate change and how it will be the poorest in the world that will bear the biggest burden.

In 2015, 193 members of the UN signed up to the SDGs (see Table 6.3). There are 17 goals ranging from poverty reduction to clean energy. There are specific targets for each SDG (UN 2017).

However, the US has some way to go if its policy outcomes are to match the rhetoric of President Obama at the UN. For example, Jeffrey Sachs, the world renowned academic and adviser to the UN on the SDGS, argues that in the US the SDGs are barely discussed or acknowledged (Mooney 2016). A recent report by Bertelsman Stiftung, a large German foundation and Sachs Sustainable Development Solutions Network has analysed the record so far of individual countries and their achievements on the SDGs. Top of the list are the four Scandinavian countries, Denmark, Finland, Norway and Sweden who are 84.5 per cent of the way to the best possible outcome across the 17 SDGs. The US, by contrast, was ranked only 25th with a 72.7 score. Its neighbour Canada was ranked 13th with a score of 76.8 per cent. The relatively poor performance of the US is the result of a number of factors according to the report. These include too many people below the poverty line, too little renewable energy and high levels of obesity (SDG Index 2016).

Away from the international stage, there were a number of domestic initiatives. In November 2015, the White House set out its strategy for federal agencies to act on climate change (White House 2015b). It stressed the need for the federal government to lead by example'. It set out new targets for federal agencies to cut greenhouse gas emissions by 41.8 per cent from 2008 levels by 2025. However, these targets were part of a broader sustainability strategy. In March 2015, Obama signed an executive order *Planning for federal sustainability in the next* decade (White House 2015b). The order sets out a comprehensive strategy including:

Table 6.3 The sustainable development goals

1	No poverty
2	Zero hunger
3	Good health and well being
4	Quality education
5	Gender equality
6	Clean water and sanitation
7	Affordable clean energy
8	Decent work and economic growth
9	Industry innovation and infrastructure
10	Reduce inequalities
11	Sustainable cities and communities
12	Responsible consumption and production
13	Climate action
14	Life below water
15	Life on land
16	Peace, justice and strong institutions
17	Partnerships for the good

- Federal agency targets for improved environmental efficiency.
- Increasing the use of renewable energy in federal agency buildings with a target of 15 per cent by 2018 rising to 30 per cent by 2025.
- A series of sustainability goals such as improving the energy efficiency of federal buildings, including energy bench marking.
- Improving water use efficiency and management, including the harvesting of storm water.
- Promoting sustainable procurement and improving recycling.
- The creation of a federal chief sustainability officer and an office of the chief sustainability officer.

There were also specific sectoral policies on sustainability. For example, in December 2016, the Obama administration launched a long term strategy on soil (White House 2016) with the aim of securing the health and sustainable use of soil, key elements in securing the future of agriculture, clear water and human health. It set out a comprehensive and detailed programme of action, including the promotion of interdisciplinary research and education and the expansion of sustainable agricultural practices.

Another example is support for bio diversity. In September 2014, the Obama administration expanded the Pacific Remote Islands Marine National Monument to encompass more than 490,000 square miles, an area six times the size when it was created in 2009. It includes seven atolls and small islands in the Pacific Ocean which are home to coral, fish, shellfish, mammals, birds, insects and plants, many unique to the region. Under the Obama administration, the whole area was put off limits to commercial resource extraction including commercial fishing. In September 2016, as his Presidency was nearing its end, Obama declared the first fully protected area in the US Atlantic Ocean, designating 4913 square miles off the New England coast as a new marine monument. The North East and Seamounts Marine National Monument lies 130 miles off the coast from Cape Cod. The area is home to many species of deep sea coral, sharks, sea turtles, sea birds and deep diving mammals such as beaked whales and sperm whales.

The Climate Reality Project (founded by former Democratic Vice President, Al Gore) names President Obama, along with Angela Merkel of Germany and Justin Trudeau of Canada as the world's top three political leaders with regard to action on climate change. It argues that 'Without his personal commitment to the issue' the historic bilateral deal with China might not have happened, a deal that helped pave the way for the Paris COP21 climate change agreement (Climate Reality Project 2016). This might well be the case. However, it has to be pointed out, as I did in Chapter 2, that for a number of commentators COP21 is too limited in its scope, and will not result in the necessary reductions in global greenhouse gas emissions required to stop runaway climate change.

The retrenchment of the Trump administration on this issue presents us with problems on two fronts. First, by threatening to remove the US from key global agreements on climate change, the Trump Presidency poses a major (if not catastrophic) threat to people and planet. Second, in relation to political strategy, the US runs the risk of losing good will, and the influence and leverage that go with it, to other political actors such as China, Germany and Canada.

Disengagement and Trump's America

The election of Donald Trump as the 45th President of the US points to a significant retrenchment on climate change and environmental protection, both domestically and internationally. Well before ascending to the Presidency, Trump's attitude to the issue of climate change and global warming were well documented. For example in 2012 in a notorious post on the social media site Twitter he said that 'The concept of global warming was created by and for the Chinese in order to make US manufacturing non-competitive'.

Since January 2017, when Trump assumed the reigns of office, there has been a whole raft of policy decisions and executive orders from the White House as part of an agenda to roll back action on climate change and environmental protection from the Obama era and beyond. These are just a few of the many examples:

- The appointment of Scott Pruit, the former attorney general of Oklahoma and a high profile climate change sceptic, as head of the EPA.
- The announcement of a 30 per cent cut to the budget of the EPA with the loss of 3200 jobs.
- All future funds for climate change research frozen.
- The removal of any reference to climate change from the official White House web site.
- The appointment of Rex Tillerson, former chief executive of Exxon-Mobil (a major funder of climate change denial lobby groups) as Secretary of State.
- Support for the highly controversial Keystone XL and Daytona Access oil pipelines, reversing the blocking of these two projects by the Obama administration.
- The easing of environmental regulation to allow coal waste to be dumped in rivers.
- Threats to pull the US out of the Paris COP21 agreement on climate change.
- Opposition from the US resulting in the dropping of a reference to financial programmes to combat climate change from the draft communiqué at G20 finance and central bankers meeting in March 2017.

In many ways, it reads more like a court indictment than a list of policy actions! All of this has led one of the world's leading climatologists, Professor Michael Mann of Pennsylvania State University, to write of the 'dizzying on going assault on science' by the Trump White House, whom he refers to in acerbic terms as the 'fossil fuel soaked administration' (Mann 2017). However, Trump's policy stance on climate change has met with opposition not just from environmentalists and climate scientists but also from key actors in the business community. One such is Jeff Immelt, CEO of General Electric (GE). In an internal company memo seen by the news site Politico, he argues that Trump's 'imagination is at work' if he does not believe in climate change, or the Paris COP21 agreement. 'We believe climate change should be addressed on a global basis through multi-national agreements' with the US playing a key role. argues Immelt. GE 'will continue to lead with our technology and actions' (Politico 2017).

Exxon Mobil, the largest American oil group, has written to the Trump administration urging it to keep the US in the Paris COP 21 agreement. Exxon Mobil argues that the Paris accord is 'an effective framework for addressing the risks of climate change' (*Millennial Journal* 2017). In a further sign of business concerns about climate change, in late November 2016, 365 heads of major US corporations sent a letter to the then President elect, Donald Trump, arguing the case that addressing climate change was vital in securing a robust US economy. The signatories included the heads of well known companies such as Hewlett Packard, IKEA, Nike and Levi Strauss. The letter was written in responses to the UN gathering in Marrakesh, Morocco, where world leaders were debating how to implement the Paris COP21 accord. Michael Kobon, the Vice President of Levi Strauss argued that 'Building an energy efficient economy in the US will ensure our nation's competiveness as leader in the global market, all the while doing the right thing for our planet' (Hanley 2016).

Conclusion

In concluding this chapter, one is reminded of the remark by Winston Churchill that *You can always count on the Americans to do the right after they have tried everything else.* Whilst Churchill might have been somewhat harsh on the US (he was always keen on the rhetorical flourish) his statement does contain the kernel of a truth. American foreign policy has always pointed in a number of directions, sometimes malevolent, sometimes a force for good, sometimes trying to do the right thing but getting it wrong. The US has the capacity to make great mistakes but it also has the capacity to reverse such mistakes, to change direction and to be a positive force. Policy on climate change is no exception to this pattern. In many ways the administration of George W. Bush marked the low point both with regard to action on climate change and environmental protection and international leadership on these matters; until the arrival of President Trump that is. Despite the criticisms levelled at President Obama, some of them well justified, his administration marked a step change from his predecessor.

In the face of extensive opposition from Congressional Republicans and Big Oil, Obama put into place a number of key policy initiatives on climate change, both domestically and internationally. Whilst trying not to overstate the point, we were given a glimpse, however slight, of the best America could be. The strengthening relationship with China on climate change, the US contribution to Paris COP21 and the global agreement in Rwanda on HFCs all pointed to an

America engaged in the global community in the fight against climate change. But this has to be balanced by a number of negatives, most notably the relative failure of the Copenhagen UNFCCC in 2009. In addition, there are the flagship trade policies TTIP and TPP which in contrast to other policy aspects of the Obama administration would have had a negative impact on climate policy and environmental protection.

The US has a responsibility to be a key player on the world stage in the fight against climate change, but such a role also brings with it a number of benefits in terms of good will, influence and social capital. China has been quick to recognise this. Indeed as Sachs has argued, the future of the US (and indeed all countries) 'lies in a healthy balance of competition and co-operation in an interconnected society' (Sachs 2011: 263).

However, the election of Donald Trump, with his America First policy, signals a reduced role for the US on the world stage, and as we have seen action on climate change is no exception to this. Mann has argued that 'we may have to withstand a vacuum in climate change leadership for the next several years' (Mann 2017), given the hold the fossil fuel lobby has on President Trump and barring any unexpected events. But as Mann argues Trump 'cannot hold back the tide of history' (Mann 2017).

References

Clinton, H. (2014) *Hard Choices*, New York: Simon and Schuster.

CLC (2017) *The Conservative Case for Carbon Dividends*, Washington, DC and London: CLC.

Cooper, R. (2015) 'The Global Fight Against Climate Change Needs a Leader. Step up America' *The Week*, 15 December.

Dernbach, J. (2009) *Sustaining America*, Washington, DC: Environmental Law Institute.

Farley, F. and Smith, Z. (2014) *Sustainability: If it's Everything is it Nothing?*, London: Routledge.

Friends of the Earth (2016) 'Investment Court System Put to the Test. New EU Proposal will Perpetuate Investors Attacks on Health and Environment' Friends of the Earth, Europe. www.foeeurope.org.

Goldenburg, S. and Harvey, F. (2014) 'Kerry: US Will Lead on Climate Change' *Guardian*, 23 September.

Guardian (2016) 'Global Deal Reached to Limit Use of HFCs' 15 October.

Jones, C. (2014) *Routes of Power: Energy and Modern America*, Cambridge, MA: Harvard University Press.

Jordan, A., Huitema, D. and van Asselt, H. (2011) 'Climate Change Policy in the European Union' in Jordan, A. Huitema, D. van Asselt, H. Rayner, T. and Berkhout, F. (eds), *Climate Change Policy in the European Union: Confronting the Dilemmas of Mitigation and Adaptation*, Cambridge: Cambridge University Press.

Mooney, C. (2016) 'A New Report Rated Countries on Sustainable Development. The US Did Horribly' *Washington Post*, 21 July.

NIC (2017) *Global Trends, Paradoxes of Progress: The Map of the Future*, Washington, DC: NIC.

Paulson, H. (2014) 'We US Conservatives Need to Recognise Climate Change and Find Market Solutions' *Observer*, 29 June.

PBL Netherlands (2015) *Trends in CO₂ Emissions: 2015 Report*, The Hague: PBL Netherlands Environment Assessment Agency.

Roberts, D. and Goldenburg, S. (2013) 'Obama and Xi Reach Greenhouse Gas Deal' *Guardian*, 10 June.

Sachs, J. (2011) *The Price of Civilization: Reawakening American Virtue and Prosperity*, New York: Random House.

SDG Index (2016) *SDG Index and Dashboard: A Global Report*, Gutersloh, Germany: Bertelsman Stiftung and Sachs Sustainable Development Solutions Network.

Sedlako, M. and Martinuzzi, A. (2012) 'Government for Sustainable Development, Evaluation and Learning: An Introduction' in Sedlako, M. and Martinuzzi, A. (eds), *Governance by Evaluation for Sustainable Development: Institutional Capacities and Learning*, Cheltenham: Edward Elgar.

UNCED (1992) *United Nations Conference on Environment and Development 1992: Rio Declaration on Environment and Development* New York: United Nations.

White House, The (2015a) 'Remarks by the President on Sustainable Development Goals' Office of the Press Secretary, 27 September, Washington DC: The White House.

White House, The (2015b) 'Obama Administration Announces 2016 Greenhouse Gas Targets and Plans' Office of the Press Secretary, 23 November, Washington DC: The White House.

White House, The (2015c) 'Executive Order: Planning for Federal Sustainability in the Next Decade' Office of the Press Secretary, 19 March, Washington DC: The White House.

White House, The (2016) 'The Obama Administration Announces New Steps to Advance Soil Sustainability' Office of the Press Secretary, 5 December Washington DC: The White House.

Internet sources

Climate Action Network (2014) Available at: http://germanwatch.org/en/download/8599.pdf, accessed 02/03/17.

Climate Reality Project (2016) 'Three Top World Leaders Fighting the Climate Crisis' www.climate-reality-project-.org/blog/three-top-world-leaders-fighting-climate-change, accessed 29/03/17.

EPA (2017) 'Draft Inventory of US Greenhouse Gas Emissions and Sinks: 1990 to 2015'. Available at: www.epa/ghgemissions/inventory-us-greenhouse-gas-emissions-and, accessed 16/02/17.

European Commission (2017) 'A Joint Assessment of the State of Play re. TTIP by EU Trade Commissioner Cecilia Malmstron and US Trade Representative, Michael Froman'. Available at: www.trade.ec.europa/doclib/press/index.cfm?id=1613, accessed 20/03/17.

Global Justice Now (2017) 'What is TTIP?'. Available at: www.globaljustice.org.uk/what-ttip, accessed 10/03/17.

Hanley, S. (2016) 'US Business Leaders to Trump: Focus on Climate Change' Available at: https://cleantechnica.com/2016/11/22/us-business-leaders-trump-focus-climate-change/, accessed 17/03/17.

Lapowsky, I. (2015) 'Tech Billionaires Team up to Take on Climate Change'. Available at:

www.wired.com/2015/11/zuckerberg-gates-climate-change-breakthrough-energy-coalition/, accessed 04/03/17.

Lorenzen, A. (2017) 'China: Global Role in Climate Change'. Available at: https://agreenerlifeagreenerworld.net/2017/01/23/evidence-mounts-that-china-wants-global-leadership-role-in-tackling-climate-change/, accessed 29/02/17.

Mann, M. (2017) 'Climate Change Denial is not Dead'. Available at: http://thehill.com/blogs/pundits-blog/energy-environment/317102-climate-denial-is-dead-long-live-climate-denial, accessed 21/03/17.

Millennial Journal (2017) 'Bishops, Catholic Leaders, Even Exxon Criticise Trump's Climate Policies'. Available at: https://millennialjournal.com/2017/03/30/bishops-catholic-leaders-even-exxon-criticize-trumps-climate-policies/, accessed 30/03/17.

Politico (2017) 'GE CEO Knocks Trump on Climate'. Available at: www.politico.com/story/2017/03/trump-climate-change-immelt-236671, accessed 30/03/17.

Politics (2016) www.politics.eu/article/obamas-next-stop-germany, accessed 15/06/16.

Scientific American (2017) 'Use it or Lose it. The Outsize Effect of US Consumption on the Environment'. Available at: www.scientificAmerican.com/article/American-consumption-habits/, accessed 30/04/17.

SPIRI (2017) Available at: www.org/research/armament-and-disaramament/arms-transfers-and-military-spending/military-expenditure, accessed 16/02/17. (N.B. The website is no longer operational).

Union of Concerned Scientists (2017) 'The US Military and Oil'. Available at: www.ucsusa.org/clean_vehicles/smart-transportation-solutions/us-military-oil-use.html#.WSQ_04WcHIV, accessed 04/04/17.

UN (2017) 'The Sustainable Development Goals'. Available at: www.un.org/sustainabledevelopment/sustainainable_development_goals, accessed 04/04/17.

Volcovici, V. and Wong, L (2016) 'China Will Soon Trump America: The Country is now the Global Leader in Climate Change Reform'. Available at: www.salon.com/2016/11/15/China-is-now-the-global-leader-in-climate-change-reform, accessed 10/12/16.

War on Want (2017) 'What is TTIP?' Available at: www.waronwant.org/what.ttip, accessed 27/03/17.

Conclusion

In a speech in June 1963 at the commencement for American University held at the John Reeves athletics centre in Washington, DC, President John F. Kennedy told his fellow Americans that' no problem of human destiny is beyond human beings. Man's reason and spirit have often solved the seemingly unsolvable and we believe that they can do it again'. Whilst climate change was far from the policy agenda back in 1963, these famous words could well apply to climate change and the challenges it presents for the US and the world in general

Barack Obama has been criticised in some circles for his rather cautious approach to action on climate change, particularly in his first term. Indeed there may be some validity in this. However, his political approach was one of trying to build a consensus and to reach across the party aisle to his political opponents. However, he faced many domestic political constraints. The challenges facing Barack Obama when he took office as the 44th President of the US in January 2009 were complex and varied. The global financial crisis that landed in the US in the autumn of 2008 (a crisis which in part helped propel Obama into the White House) was sweeping across the US economy, with a devastating impact on jobs and the housing market. Immediate action in the form of a financial stimulus bill was the first priority to prevent an economic slump. As a consequence, action on climate change took something of a back seat in the early period of the Obama administration.

Yet in assessing his first term in office, Klyza and Sousa argue that Obama's actions on climate change did make a difference (Klyza and Sousa 2013: 303). Indeed as we have seen the Obama administration's engagement with action on climate change marked a significant departure from the obstructionist policies of the George W. Bush era. However, one cannot overstate the rancorous nature of the political debate that greeted Obama on his election to the Presidency. From the day he took office, large swathes of Congressional Republicans, especially those linked to the Tea Party movement, were determined to thwart him. Backed by big oil they sought at every turn to block action on climate change.

As a consequence, little if any legislation on climate change was passed by Congress during Obama's first term in office. The failure of Congress to pass a carbon cap and trade bill, a policy supported by Obama, marked the effective end of attempts to legislate on climate change through Congress. Instead, the

President had to rely on a series of executive orders and actions to push through his policy agenda. Among such policies was the 2009 New Energy for America Plan and the 2011 Blue Print for a Secure Energy Future.

In his second term in office, Obama restated his commitment to tackle climate change, and made clear to Congress his determination to use his executive authority if Congress was unable or unwilling to act. There followed a number of policy initiatives. These included the CPP, arguably one of the key environmental policy initiatives since the 'golden age'. On the international arena, the growing US/China relationship on climate change played a key role in the COP21 Paris climate change agreement in December 2015. Despite its limitations, COP21 represented a significant development in international co-operation on climate change. In September 2016, as his Presidential term was coming to an end, Obama announced that he would seek to play an active role on both the US and world stage, seeking to raise the profile of action on climate change.

In January 2017, Donald Trump settled into the White House as the 45th President of the US. His election signals an attempt to reduce action on climate change. Trump has issued a string of executive actions to roll back action on climate change, including the rescinding of the CPP. He has also hinted that he may take the US out of the COP21 agreement. Yet such actions are set to be the subject of protracted legal wrangles and increasing opposition. States and cities that are taking action on climate change will continue to do so. For example, in March 2017, 75 US mayors, from the Mayors National Climate Action Agenda came out in direct opposition to Trump's executive order on the CPP, stating that they would continue their collective work to take action on climate change. Also we have also seen how many businesses will continue to invest in clean energy as part of their ongoing commercial planning. So action on climate change will continue despite the best efforts of the Trump administration.

Mann has argued that 'we have to withstand a vacuum in climate change leadership for several years' (Mann 2017) given the hold the fossil fuel lobby has on President Trump, barring any unexpected events. But as Mann argues, Trump 'cannot hold back the tide of history' (Mann 2017). But in the final analysis, the threat of climate change to the US and the world is real and substantive. There are no alternative facts to this. The Obama administration recognised this. No matter how the winds of politics change, the reality will not change and the politics will have to deal with it.

References

Klyza, C. and Sousa, D. (2013) *American Environmental Policy Beyond Gridlock* Cambridge, MA: MIT Press.

Mann, M. (2017) 'Climate Change Denial is not Dead'. Available at: http://thehill.com/blogs/pundits-blog/energy-environment/317102-climate-denial-is-dead-long-live-climate-denial, accessed 21/03/17.

Index

Page numbers in **bold** indicate figures and in *italic* indicate tables, end of chapter notes are indicated by a letter n between page number and note number.